SOLID ROCKET PROPULSION FOR SPACE EXPLORATION

by

Dr. Ugur Guven

Aerospace & Nuclear Engineer (PhD)

Gurunadh Velidi

Mechanical & Energy Systems Engineer

This book was written by using compilation of various thesis reports to create a book that adressed the various issues in solid rocket propulsion. We thank our students and other colleagues who have helped in accumulating the information.

Prof. Dr. Ugur GUVEN
Aerospace & Nuclear Engineer (PhD)

Gurunadh Velidi
Mechanical & Energy Systems Engineer

TABLE OF CONTENTS

                                                  Page

TABLE OF CONTENTS 3
ABBREVIATIONS 5
LIST OF TABLES 6
LIST OF FIGURES 7
SUMMARY 8
1. INTRODUCTION 9
   1.1 Current Day 10
   1.2 Gravity 10
   1.3 Propulsion 11
   1.4 Types of Propulsion 12
      1.4.1 Chemical Rocket Propulsion 12
      1.4.2 Electric Propulsion 13
      1.4.3 Nuclear Propulsion 13
      1.4.4 Solar Thermal Propulsion 13
   1.5 Other forms of Propulsion 13
   1.6 Types of Chemical Rocket Propellants 14
      1.6.1 Solid Propellants 14
      1.6.2 Liquid Propellants 14
      1.6.3 Gelled Propellants 15
      1.6.4 Hybrid Propellants 15
      1.6.5 Cold Gas Propellants 15
   1.7 Principle of Operation 16
   1.8 Newton's Third Law and The Rocket Equation 16
   1.8.1 Tsiolkovsky's Rocket Equation 17
      1.8.2 Total Impulse 18
      1.8.3 Specific Impulse 19
      1.8.4 Mass Ratio 19
      1.8.5 Thrust 20
2. SOLID ROCKET PROPULSION 22
   2.1 Solid Propellants 22
      2.1.1 Black Powder Propellants 24
      2.1.2 Zinc-Sulphur Propellants 24
      2.1.3 Candy Propellants 24
      2.1.4 Double-Base Propellants 24
      2.1.5 Composite Propellants 25
      2.1.6 High Energy Composite Propellants 25
      2.1.7 Composite Modified Double-Base Propellants 25
      2.1.8 Smokeless Propellants 25
   2.2 Parameters for Formulating Propellants 26
   2.3 Chemical Kinetics 29
3. SOLID FUELLED ROCKET ENGINE 31
   3.1 Basic Configuration 32
   3.2 Design of Solid Fuelled Rocket Engine 33

    3.2.1 Sold Fuelled Rocket Engine Case 34
    3.2.2 Propellant Grain Design 34
    3.2.3 Propellant Grain Profile 36
    3.2.4 Combustion Chamber 38
    3.2.5 Insulation 38
    3.2.6 Igniter 39
    3.2.7 Nozzle 41
3.3 Thrust Vector Control System 43
    3.3.1 Secondary Fuel Injection 44
    3.3.2 Mechanical Deflectors 44
    3.3.3 Movable Nozzle TVC 44
    3.3.4 Flexbearing 45
    3.3.5 Ball-and-Socket Joint 45
    3.3.6 Fluid Bearing 45
    3.3.7 Hinged Movable Nozzle 45
    3.3.8 Gimbaled Nozzle 45
    3.3.9 Rotatable Nozzle 46

4. THERMODYNAMICS OF SOLID ROCKET PROPULSION 48
4.1 Laws of Thermodynamics 49
    4.1.1 The Zeroth-law of Thermodynamics 49
    4.1.2 The First law of Thermodynamics 49
    4.1.3 The Second law of Thermodynamics 50
4.2 Nozzle Flow Equation 54
4.3 Rocket Nozzle Performance 55
4.4 Effect of Area Variationon on Flow Properties 57

5. SPACE SHUTTLE'S SOLID ROCKET BOOSTERS 59
5.1 Solid Rocket Booster Structural Design 60
5.2 Solid Rocket Booster Propellant Ignition 61
5.3 Solid Rocket Booster Propellant Features 62
5.4 Solid Rocket Boosters Seperation 63
5.5 Solid Rocket Boosters Reuse and Recovery 64
5.6 Summary 64

6. INVISCID AND COMPRESSIBLE FLOW THROUGH A CONVERGING DIVERGING NOZZLE 65
6.1 Problem Description 66
6.2 Setup and Solution for Viscous Flow 66
6.3 Setup and Solution for Inviscid Flow 72
6.4 Results 76
6.5 Recommendation 76

REFERENCES 77

ABBREVIATIONS

FAI	: Federation Aeronautic Limited
LEO	: Low Earth Orbit
ANCP	: Ammonium Nitrate Composite Propellant
APCP	: Ammonium per Chlorate Composite Propellant
HTPB	: Hydroxyl-Terminated Polybutadiene
CL-20	: China Lake Compound#20
AP	: Ammonium Perchlorate
AN	: Ammonium Nitrate
ADN	: Ammonium Dinitramide
HNF	: HydraziniumNitroformate
RDX	: Cyclotrimethylenetrinitramine
HMX	: Cyclotetramethylenetetramintramine
HTPB	: Hydroxyl-Terminated Polybutadiene
CTPB	: Carboxyl-Terminated Polybutadiene
PBAN	: Polybutadiene Acrylonitrile
PEG	: Polyethylene Glycol
PPG	: Polypropylene Glycol
NC	: Nitrocellulose
GAP	: GlycidylAzide Polymer
DOA	: DioctylAdipate
DOP	: Dioctyl Phthalate
NG	: Nitroglycerin
BTTN	: ButanetriolTrinitrate
TMETN	: Trimethyloethanetrinitrate
MNA	: P-Rc-Methyl Nitroanaline
TVC	: Thrust Vector Control
ET	: Exposure Time
MAR	: Material Affected or Char Rate
SF	: Safety Factor
TP	: Thermal Protection Thickness
MT	: Manufacturing Tolerance
SRBs	: Solid Rocket Boosters
SSMEs	: Space Shuttle's Main Engines
ET	: External Tank

LIST OF TABLES

Page

Table 2.1: Common propellants materials and their functions..................................27
Table 3.1: Thrust vector control mechanism. 46

LIST OF FIGURES

Page

Figure 1.1 : How a rocket engine works. 16
Figure 1.2 : Space shuttle launch 17
Figure 1.3 : Tsiokolvky's Rocket equation. 20
Figure 1.4 : Exhaust parameters of rocket nozzle 21
Figure 3.1 : Solid fuelled rocket engine. 32
Figure 3.2 : Solid fuelled rocket engine design. 33
Figure 3.3 : Features used in propellant grain design 35
Figure 3.4 : Cross sections of propellant grain 37
Figure 3.5 : Insulating layers for thermal protection 39
Figure 3.6 : Ignition and grain heating of a solid propellant rocket motor 40
Figure 3.7 : A convergent-divergent nozzle (de-Laval Nozzle) 42
Figure 3.8 : Flow process inside the nozzle 42
Figure 3.9 : Thrust vector control system mechanism 43
Figure 4.1 : Flow through a de Laval Nozzle 52
Figure 4.2 : Effect of temperature, pressure, velocity inside the nozzle 57
Figure 5.1 : Space shuttle launch profile 60
Figure 5.2 : Solid rocket booster structural design 61
Figure 5.3 : Solid rocket booster propellant ignition 62
Figure 5.4 : Solid rocket booster propellant features 63
Figure 5.5 : Solid rocket booster's separation 63
Figure 6.1 : Problem Schematic 66
Figure 6.2 : Grid Display 67
Figure 6.3 : Scaled Residuals 69
Figure 6.4 : Contours of static pressure 70
Figure 6.5 : Contours of Mach number 70
Figure 6.6 : Contours of static temperature 70
Figure 6.7 : Velocity vectors 71
Figure 6.8 : Magnified view of velocity vectors 71
Figure 6.9 : Scaled Residuals (Inviscid Flow) 73
Figure 6.10 : Contours of static pressure (Inviscid Flow) 74
Figure 6.11 : Contours of Mach number (Inviscid Flow) 74
Figure 6.12 : Contours of static temperature (Inviscid Flow) 74
Figure 6.13 : Velocity vectors (Inviscid Flow) 74
Figure 6.14 : Magnified view of velocity vectors (Inviscid Flow) 75

ANALYSIS OF SOLID-FUELLED ROCKET PROPULSION

SUMMARY

The objectives of this book is to present the fundamentals of Solid Rocket Motor, starting from the elementary analysis of rockret propulsion and then justifying the need of sophisticated computation of the internal flow. After a brief reminder of solid rocket theory, a description of its main components is proposed. The elementary parameters controlling the operation are introduced and the basic formula predicting the steady-state operation pressure is established. In this book, we have described about solid solid rocket propulsion and we explored some of the issues which are related to the performance of the whole rocket. We have also described the main design of the rocket engine and its performance related factors and design aspects which affect the perfomance of the motor , and practical limitations for motor design. The main issues faced by the Solid Rocket Engine require an accurate description of internal aerodynamics, either to predict the pressure/thrust programs and the normal tansient phase like ignition, or to motor stability. A short overview of the evolution of the Solid Rocket Motor internal aerodynamics during the last thirty years is also given in the book. It is hoped that this book will provide an introductory substance to the field of solid rocket propulsion.

1. INTRODUCTION

The Earth is just too small and fragile a basket for the human race to keep all its eggs in.'

The development of a country's people has always been closely related with transportation. There is no clear path between the Space and the Earth's atmosphere. As we move away towards the Earth's atmosphere, the air density and pressure changes in different atmospheric layers. Karman line was defined as the boundary layer between the Earth's atmosphere and Space in the middle of 1950s. It was fixed at an altitude of 100 kilometres. The Federation Aeronautic International (FAI) established this line, commonly known as the world air sports federation. At the Karman line boundary there is still a perceptible drag caused by the atmosphere.

At the beginning of 21st century, only few humans travelled in rocket propelled vehicles due to this the rocket may be seen as the upcoming social order in transport. A surprising amount of domestic and commercial communication is now depending upon satellites. Most of our information related to the country, medical, security, government activities etc. travels from one part of the Earth to another, from the internet, through news images, telephone calls and space observations. To explore the mystery of the space, rocket propulsion is the necessary transportation technology and it will help in the growth of human communication and generate interest to learn more about space and it's mysteries. It affects the lives and work of a growing number of people, who want to understand the principles behind the rocket technology and its technical details.

1.1 Current Day

Rockets are the popular military weapons of today and key component of the space technologies. Rockets are generally used in large battlefields, V-2 types has given way to guided missiles. Due to its long range capability and accurate aiming, rockets are often used by helicopters and light aircraft for ground attack. Rockets are several times more devastatingthan automatic machine guns.

Economically, rockets enabled all of the space technologies particularly satellites. The satellites impact people's everyday lives in almost countless ways. Scientifically, it has opened a window in the mysteries of the universe. It explores the solar system and the space-based telescope that obtained a clearer view of the rest of the universe by the launch of space probes which was possible only with the help of rocket technology. Though, it is may be manned spaceflight that has mostly caught the thoughts of the public. Vehicles such as Space Shuttle for scientific research, the Soyuz gradually more for orbital tourism and Spaceship One for suborbital tourism may demonstrate a trend headed for greater commercialisation of manned rocketry. Chemical rockets will almost certainly make an important role in space operations for the future. Even though highly developed, high-exhaust velocity rockets will finally take over most long distance missions; chemical rocket and nuclear-thermal rockets will remain the recognised leaders in producing the high thrust-to-weight ratio essential for leaving most planetary surfaces. (Turner, 2009)

1.2 Gravity

Since the advent of the space age, the main concern has been to overcome gravity, which has been an important obstruction to leaving Earth. In essence, this has hampered space investigations more then anything as many of the rocket designs depend on the availability of the rockets to escape Earth's gravity well.

Gravity causes all the objects to attract each other. On the Earth the effect of gravity can be seen. Gravity acts along a line between the centres of the objects. On the Earth, objects are always attracted towards the centre of the Earth or, as seen from the surface, downward. The force acts between the object and the Earth, is called the gravitational force. The size of the force depends on the object and the distance between them. A larger mass produces a greater force. Just as the mass of the Earth produces gravitational attraction, so the other planets and the Sun also produce a gravitational attraction relative to their masses. The size of the force of attraction from an object decreases rapidly with distance.

To overcome gravity, energy is required. In order to escape Earth's atmospheric effects, we have to achieve the velocity, called the escape velocity. From the surface of the Earth this value is about 11.2 kilometres per second or about 40,820 kilometres per hour, however, the value decreases with distance away from the surface. When a space vehicle or rocket approaches another body, such as the moon or a planet, there comes a time when the gravity from that body attracts it more than the Earth attracts it. At this stage, effort is no longer required to escape from the Earth, but effort would now be required to get back to the Earth. It is possible to overcome gravity and travel in any direction in space. However, this requires continual thrust from a rocket, which requires a lot of power.

1.3 Propulsion

On the Earth, the forward motion is usually achieved by pushing on some medium, such as the ground for a car and the sea for a motor boat. Although most propulsion systems do push on something, the act of throwing something out in the opposite direction can also produce forward motion. These methods of propulsion use an action that causes a reaction, which was described in 1687 by Sir Isaac Newton and his third law of motioncalled "For every action there is an equal and opposite reaction". High up in the atmosphere the molecules of air are apart, and at altitude greater than about 80 kilometres the atmosphere no longer exists as an effective medium as it is almost a vacuum. At these altitudes as it requires the oxygen in the air to act as an oxidiser, which enables the fuel to burn. A reaction motor that carries all of its propellant and, if required, the fuel and oxidiser with it, can overcome an absence of air. To remain in space and not fall back to the Earth, a spacecraft must either go into an orbit around the Earth or be propelled so fast that it reaches or exceeds the escape velocity and the thrust away from the Earth is larger than the pull of gravity. (Rogers, 2008)

1.4 Types of Propulsion

    1.4.1   Chemical Rocket Propulsion
    1.4.2   Electric Propulsion
    1.4.3   Nuclear Propulsion
    1.4.4   Solar Thermal Propulsion

1.4.1 Chemical Rocket Propulsion

Chemical rocket propulsion is the only method we currently have to escape from the surface of the Earth's atmosphere in most of space shuttle, satellite launch etc. Chemical rocket propulsion system can produce enough thrust to overcome the gravity on the Earth's surface. Chemical rockets not only work in space, but they also work in the atmosphere. Chemical rocket propulsion systems uses chemical propellants to produce large amount of thrust. At the time of burning chemical propellants can be solid, liquid or gas or sometimes they can be used in combination with each other. When two different types of propellants are used, the system is called the bi-propellant system.

It is neccassary to use the combination of a fuel and oxidizer to burn the chemical propellants. The oxidizer is the useful component that contains all of the oxygen required to allow the chemicals to react or burn and the presence of oxidizer needs no external oxygen supply.

1.4.2 Electric Propulsion

According to the method of accelerating the propellant, electric propulsion can be classified in three different manner. These are electro-thermal, electrostatic and the electromagnetic. These three types of electric propulsion systems requires an electric power source, but the electric power source is usually separate to the propulsion system.

The thrust generated from an electric propulsion system is usually low, about 0.005-1 Newton and due to this only a small acceleration is attained and has a very high specific impulse rating. Although this acceleration can be sustained for a period of weeks or months.

1.4.3 Nuclear Propulsion

At the beginning of the twentieth century the idea of the use of nuclear energy in rocket propulsion system adopted by many satellite, space shuttle launch agency. In nuclear propulsion nuclear powered rockets use nuclear energy and this energy superheats the hydrogen gas to a temperature at which the exit velocity of the rocket is very high. The main problem concern with this design is that the reactor of the rockets tend to outlast of hydrogen gas. The concept is possible, but debate continues as to the safety of nuclear power using in space, just as it does on the safety of terrestrial nuclear power plants.

1.4.4 Solar Thermal Propulsion

Propulsion systems usually requires combustion or electric power source that heat and expand propellants. In solar thermal propulsion technology, the heat from the Sun could be used directly. The STP system would be naturally by focusing the Sun light onto a chamber containing the propellant. However, no solar thermal propulsion system has yet been used in space. The benefit of the solar thermal propulsion being a lighter system does not yet overcome this disadvantage.

1.5 Other Forms of Propulsion

Other forms of propulsion such as solar sails, which use sunlight to propel the craft have been suggested and these methods have become technically and financially viable, but others types of propulsion systems, such as space elevators, that used solid link between the Earth and a geostationary orbit, remain, for now at least, in the realms of science fiction due to technical constrains. (Rogers, 2008)

1.6 Types of Chemical Rocket Propellants

    1.6.1   Solid Propellants

    1.6.2   Liquid Propellants

    1.6.3   Gelled Propellants

    1.6.4   Hybrid Propellants

    1.6.5   Cold Gas Propellants

1.6.1 Solid Propellants

Solid Propellants are the simplest type of rocket propulsion system. A firework rocket is use the similar concept in design to a rocket motor. Solid propellants are used in the main propulsion system for small and medium launchers application. Solid propellants (and almost all rocket propellants) consist of an oxidiser and a fuel. The solid propellant is storable, and is safe to handle for safety purpose. Solid propellants requires no propellant delivery system, and this produce a huge improvement of consistently good quality and cost. The standard high energy solid rocket fuel consist Ammonium Perchlorate Composite material etc.

1.6.2 Liquid Propellants

Liquid Propellant engines are less reliable and more complex than solid rocket motors. The propellant is insert to the combustion chamber by static pressure in the tanks. High pressure gas is generated by evaporation of the propellant and introduced to the tanks, and in this way forces the fuel and oxidiser into the combustion chamber. The main advantage of liquid propellant engines that it can be control to start and stop the engine and to restart it if required. They may also contain controls to drain and fill the propellant and various types of safety devices that self-check the proper functioning of the systems.

1.6.3 Gelled Propellants

Some liquid propellants have additives that are in the form of jelly when they are still, but return to a fluid when excited. The gelled propellant can flow through pumps, pipes and valves when an sufficient force is applied to it but to a solidified gel when there is no force. The loading or unloading of gelled propellants is more difficult than for normal liquid propellants also have a little-bit lower specific impulse, which makes it less efficient than liquid propellants.

1.6.4 HybridPropellants

A hybrid propulsion system is the combination of liquid and solid propellant where one propellant is kept in a solid phase and the other as a liquid. Most frequently the fuel is a solid and the oxidizer is a liquid. The highly pressurized oxidiser is inserted through the grain and turns it into vaporize form and the exhaust enters in a mixing chamber to ensure complete combustion before exiting through nozzle. The main advantage of this type of system is safety during operation, storage, fabrication, stop, start and restart capabilities. Hybrid propellants have the higher specific impulse than solid rocket motors.

1.6.5 Cold Gas Propellants

Cold gas propellants are very simple propellant system, that uses a high pressure gas tank, valves. In cold gas propellant systems the escaping gas is controlled and directed through the suitable nozzle to produce the required action and reaction. In cold gas propellant system high density gases such as nitrogen, argon, and krypton are used. (Sutton, 2001)

1.7 Principle of Operation

Rocket engines produced thrust by the exhaust of a high-speed gas flow velocity. The pressure generated by the highly compressed gases of solid or liquid propellant is nearly about 10-200 bars inside the combustion chamber of solid or liquid propellants, consisting of fuel and oxidiser components. The exhaust of these combustion products is then passed through a propelling nozzle which generally uses the heat energy of the gas and accelerate the exhaust to very high speed, move the engine in the opposite direction.

In rocket engines, the performance of rocket engine mainly depends upon the high temperatures and pressures as this permits a longer nozzle to be fitted to the engine, which gives higher exhaust velocity, as well as giving better thermodynamic efficiency.

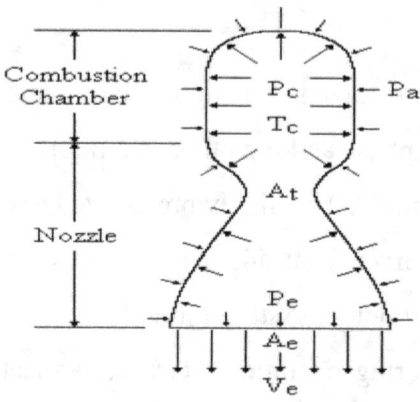

Figure 1.1: How a rocket engine works

1.8 Newton's Third Law and The Rocket Equation

A rocket is a device which propels itself by emitting high speed exhaust jet. Generally most of the rocket engines are known as internal combustion heat engines. A rocket engine is a jet engine that produces high speed jet velocity. This high speed jet velocity produced by combusting a solid, liquid or gaseous fuel with an oxidiser within a combustion chamber that results the extremely hot compressed gases escape through high expansion ratio nozzle and produces highest specific power and highest specific thrust. Rocket engines are also called the reaction engines. The basic principle behind the working the working of rocket engine is "Newton's third law".

- Force equals mass times acceleration, $F = ma$; or force equals the change of momentum with respect to time, $F = \dfrac{dp}{dt}$.
- The principle of inertia: A body in motion (or at rest) tends to remain in motion (or at rest) unless acted on by a force.
- Every action produces an equal and opposite reaction.

The momentum carried away by the jet results in force, acting so as to accelerate the rocket in the direction opposite to that of jet. The essential facts are that the rocket accelerates, and its mass decrease. (Turner, 2009)

Figure 1.2: Space shuttle launch

1.8.1 Tsiolkovsky rocket equation

Konstantin Tsiolkovsky (1875-1935), a mathematics teacher, wrote about space travel, including escape velocity and weightlessness, in 1883, and he wrote about the artificial satellites in 1895. Tsiokolvsky derived the rocket equation, and study in detail with the use of rocket propulsion for space travel outside the Earth's atmosphere and he also explain multistage rocket propulsion. At the time of study about rocket propulsion, he find that the exhaust velocity is the important performance parameter for rocket propulsion.

The force that projects the exhaust and the force that propels the rocket are same in magnitude and reacts in opposite direction relative to each other. It partakes in

Newton's third law, "action and reaction are equal and opposite", where action means force. The accelerating force is represented, using Newton's law, as

$$F = m.a = m.\frac{dv}{dt} \quad (1.1)$$

$$F = \dot{m}.V_e = -\frac{dm}{dt}.V_e \quad (1.2)$$

In this equation, the thrust of the rocket is expressed in terms of the mass flow rate $(\dot{m})$, and the effective exhaust velocity, V_e. The effective exhaust velocity (V_e) is the average equivalent velocity at which propellant is ejected from the vehicle.

Thus by equating these two equations

$$m\frac{dv}{dt} = -\frac{dm}{dt}.V_e \quad (1.3)$$

The resultant formula which Tsiolkovsky obtained for the vehicle velocity V is

$$V = V_e.\log_e\frac{M_o}{M} \quad (1.4)$$

Here M_o is the mass of the rocket at ignition, and M is the current mass of the rocket. This simple formula is the basis of all rocket propulsion. The velocity increases with time as the propellant is burned. It depends upon natural logarithm of the ratio of initial mass to current mass; that is, on how much of the propellant have been burned. For a fixed amount of propellant burned, it also depends on the exhaust velocity. (Turner, 2009)

1.8.2 Total Impulse

The total impulse I_t is the thrust force integrated over the burning time t.

$$I_t = \int_0^t F.dt \quad (1.5)$$

For constant thrust and negligible start and stop transients this reduces to

$$I_t = F.t \quad (1.6)$$

I_t is proportional to the total energy released by all the propellant in a propulsion system.

1.8.3 Specific Impulse

The specific impulsion I_{sp} is the total impulse per unit weight of propellant. It is the important figure as the performance of rocket propulsion system. A higher specific impulse means high performance of rocket. If the total mass flow rate propellant is $\dot{m}$ and the standard acceleration of gravity at sea level g_o is *9.8066 m/sec²* or *32.174 ft/sec²*, then

$$I_{sp} = \frac{\int_0^t F\, dt}{g_o \int \dot{m}\, dt} \tag{1.7}$$

This equation will give a time-averaged specific impulse value for any rocket propulsion system, particularly where the thrust varies with time. For constant thrust and propellant flow this equation can be simplified; below,

$$I_{sp} = \frac{I_t}{m_p g_o} \tag{1.8}$$

Where m_p is the total effective propellant mass. The product $m_p g_o$ is the total effective propellant weight w. For constant propellant mass flow rate $\dot{m}$, constant thrust F, and negligibly short start or stop transients:

$$I_{sp} = \frac{F}{\dot{m} g_o} = \frac{F}{\dot{w}} \tag{1.9}$$

Where $\dot{w}$ is the weight flow rate of the solid fuelled rocket engine. The unit of specific impulse is *lbf/sec*. The numerical value of I_{sp} is the same in the *EE* and the *SI* system of units.

1.8.4 Mass ratio

The mass ratio is just the ratio of the initial mass to the current mass. It is written as *R*, or *Λ*.

$$R = \frac{M_o}{M} \tag{1.10}$$

As shown in figure (1.3), the rocket velocity is plotted as a function of the mass ratio. The rocket equation shows that the final velocity depends upon only two numbers: the final mass ratio, and the exhaust velocity. It does not depend on the thrust, or the size of the rocket engine, or the time rocket burns, or any other parameter. Clearly, a higher exhaust velocity produces a higher rocket velocity, and much of the effort in rocket design goes into increasing the exhaust velocity. The point at which the rocket speed exceeds the exhaust speed is when the mass ratio becomes equal to *e*, or *2.718*, the base of natural logarithms. (Turner, 2009)

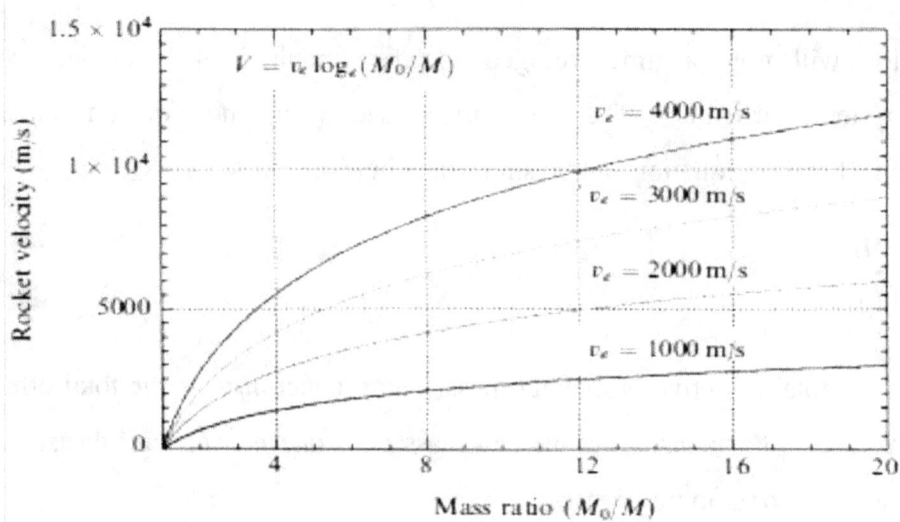

Figure 1.3: Tsiokolvsky's rocket equation (rocket velocity vs mass ratio)

1.8.5 Thrust

The thrust is the amount of force an engine produces on the rocket. The amount of thrust, along with the rocket mass, determines the acceleration. The mission profile will determine the required and acceptable accelerations and thus, the required thrust. The amount of thrust produced by the rocket depends on the mass flow rate through the engine, the exit velocity of the exhaust, and the pressure at the nozzle exit. All of these variables depend on the design of the nozzle. If the free stream pressure and pressure at the exit of the nozzle is given by P_o and P_e respectively, the thrust equation becomes:

$$F = \dot{m}.V_e + (P_e - P_o).A_e \tag{1.11}$$

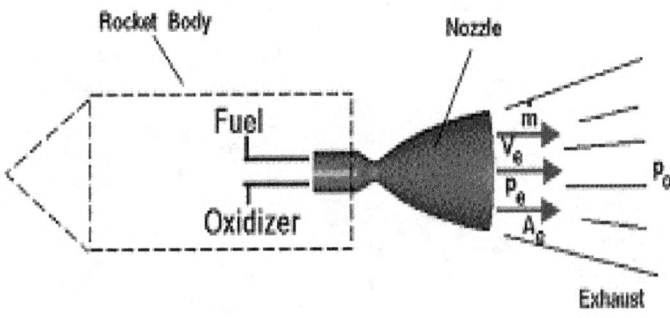

Figure 1.4: Exhaust parameters of rocket nozzle

Thethrust equation shown above, works for both solid rocket engines and liquid rocket engines. The first term shown in the above equation is the momentum thrust represented by the product of the propellant mass flow rate and its exhaust velocity relative to the vehicle. The second term represents the pressure thrust consisting of the product of the cross-sectional area at the nozzle exit A_e.

2. SOLID ROCKET PROPULSION

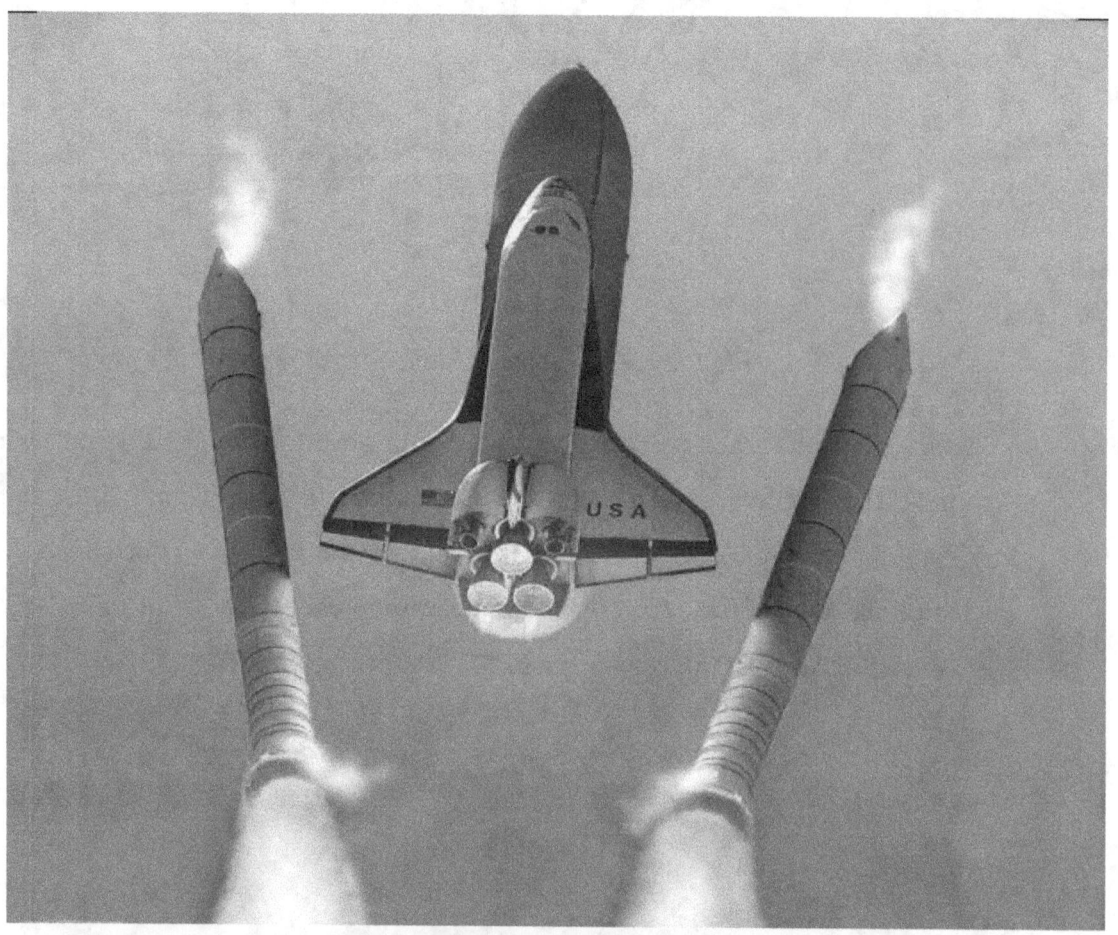

"There are a thousand things that can happen when you go light a rocket engine, and only one of them is good."

Solid rocket propulsion is the begining of the propulsion; however it is not often used in aeronautics outside its use as booster rockets. The Solid rocket Propulsion is the form chemical rocket propulsion system in which solid propellants are used. Solid rocket propulsion system is the simplest form of all rocket propulsion system designs. Solid rocket propulsion system is used as light launch vehicles for Low Earth Orbit (LEO) payloads under 2 tons or escape payloads up to 1000 pounds. The propellant to be burned contained within the combustion chamber or case which contain all the chemical constituents (fuel plus oxidiser) for complete burning. When ignited, the propellant compounds burn instantly, expelling hot gases from a nozzle to produce the high exhaust thrust. Solid propellant motors are not like liquid propellant motors because they can't be shut down. There are two main disadvantages of this system: (Sutton, 2001)

- Once ignited, they burn until all the propellant is exhausted, because it contains all the ingredients necessary for combustion within the chamber in which they are burned.
- The specific impulse is rather low because of the low chemical energy of the solid propellant.

2.1 Solid Propellants

Solid rocket can supply high thrust for relatively short period of time. Propellant combinations are usually to some extent higher for solid propellant first stages than for upper stages. For this reason, solid propellants have been used as initial stages in rockets, while reserving high specific impulse engines. A solid rocket propellant may be filled with as 91% of solid particles that consists of an elastomeric polymer. The solid propellant is formulated using a combination of a fuel and an oxidizer may have 4 to 12 ingredients that are intimately mixed on a microscopic level, and in some cases, are parts of the same molecule. Initially, the ingredients of the solid propellant are mixed together to form a thick liquid containing suspended solid particles. Once the propellant has been insert into the motor case, the mixture hardens enough to sustain its mechanical integrity and maintain sufficient elasticity to decrease the happening of cracking as it experiences induced stresses from operation (high pressure) and storage (thermal expansion and contraction). Modern propellants can be categorised in several ways, as described below. Sometimes the same propellant will fit into two or more of the classifications.

2.1.1 Black powder propellants

Black powder propellant are the composition of charcoal (fuel), potassium nitrate (oxidizer), and sulphur (additive), it is one of the oldest pyrotechnic compositions with application to rocket propulsion. Now a days, black powder finds use in low-power model rockets, as it is cheap and fairly easy to produce. The fuel grain is typically a mixture of compressed fine powder (into a solid, hard slug), with a burn rate that is highly dependent upon operating conditions and exact composition and operating. Black powder is very sensitive to fracture (sudden failure upon ignition) and poor performance, black powder does not typically find in use in motors above 40 Ns.

2.1.2 Zinc-Sulphur Propellants

Zinc-sulphur propellantsare the composition of powdered zinc metal (fuel), and powdered sulphur (oxidizer) is another compressed propellant that does not find any practical application outside of specialized limited skill rocketry circles due to its poor performance (as most zinc-sulphur burns outside the combustion chamber) and incredibly fast linear burns rate on the order of 2 m/s, leaving a spectacular large orange fireball behind it.

2.1.3 Candy propellants

In general, candy propellants are the composition of an oxidizer (characteristically potassium nitrate) and a sugar fuel (characteristically dextrose, sorbitol, or sucrose). These types of propellants are introduce into shape by appropriate melting the propellant comprises together and flowing or padding the amorphous colloid into a mold. Candy propellant induces a low-medium specific impulse of roughly 130 s and, therefore, are used chiefly only by non-professional and experimental rockets.

2.1.4 Double-Base Propellants

Double-Base propellants are the composition of two types of monopropellant fuel ingredient. One of them characteristically act like as a extreme-energy monopropellant and the second type acts as a lower-energy steady monopropellant. In typical conditions, nitro-glycerine is crumble in a nitrocellulose gel and solidified with additives. Double-Base propellants are executed in applications where least smoke is required yet medium-high functioning (I_{sp} of roughly 235 s) is required. The contribution of metal fuels (such as aluminium) can increase the functioning, however metal oxide nucleation in the exhaust can convert the smoke impossible to see direct.

2.1.5 Composite Propellants

A powdered oxidizer and powdered metal fuel are intimately mixed and immobilized with a rubbery binder (that also acts as a fuel). Composite propellants are often either ammonium nitrate-based (ANCP) or ammonium per chlorate-based (APCP). Ammonium nitrate composite propellant often uses magnesium and/or aluminium as fuel and delivers medium performance whereas Ammonium per chlorate composite propellant often uses aluminium fuel and delivers high performance. Composite propellants are cast, and retain their shape after the rubber binder, such as Hydroxyl-terminated polybutadiene (HTPB), cross-links (solidifies) with the aid of a curative additives. Because of its high performance, moderate ease of manufacturing, and moderate cost, APCP finds widespread use in space rockets, military rockets, amateur and hobby rockets. Ammonium di-nitramide, $NH_4N(NO_2)_2$, is being considered as a 1-to-1 chlorine-free substitute for ammonium per chlorate in composite propellants.

2.1.6 High-Energy Composite Propellants

Typical high energy composite propellants start with a standard composite propellant mixture (such as APCP) and add a high energy explosive to the mix. This extra component usually is in the form of small crystals of RDX or HMX, both of which have higher energy than ammonium per chlorate. Despite a modest increase in specific impulse, implementation is limited due to the increased hazards of the high-explosive additives.

2.1.7 Composite Modified Double Base Propellants

Composite modified double base propellants start with a nitrocellulose/nitro-glycerine double base propellant as a binder and add solids (typically ammonium-per-chlorate and powdered aluminum) normally used in composite propellants. The ammonium-per-chlorate makes up the oxygen deficit introduced by using nitrocellulose, improving the overall specific impulse. The aluminum also improves specific impulse as well as combustion stability. The mixing of composite and double base propellant ingredients has become so common as to blur the the functional definition of double base propellants.

2.1.8 Minimum-signature (smokeless) propellants

One of the most active areas of solid propellant research is the development of high energy, minimum-signature propellant using CL-20 (China Lake compound#20), $C_6H_6N_6(NO_2)_6$, which has 14% higher energy per mass and 20% higher energy density than HMX. The new propellant has been successfully developed and tested in tactical motors. The propellant is non-polluting: acid free, solid particulates free, and lead free. It is also smoke free and has only a faint shock diamond pattern that is visible in the otherwise transparent exhaust. Without the bright flame and dense smoke trail produced by the burning of aluminized propellants, these smokeless propellants all but eliminate the risk of giving away the positions from which the missiles are fired. CL-20 is considered a major breakthrough in solid rocket propellant technology but has yet to see widespread use because costs remain high. (Sutton, 2001)

2.2 Parameters for Formulating Propellants

Many interacting factors must be conceived when developing a solid rocket propellant. A fundamental consideration is for the propellant to put up sufficient energy toreceivemission requirements. In contribution to specific impulse (I_{sp}), a evaluation of the amount of thrust the propellant provides per unit mass, solid rocket motors are generally volume-limited. Therefore propellant density must be estimated. The working pressure and mass flow rate of a solid rocket motor relying on the ballistic properties of the propellant, hence these properties must be steadily controlled. Propellants generallyuse very energetic materials to obtain the performance designs of the rocket motor. They may be shock-sensitive, poisonous to react explosively when destructed and ignited. Each of these considerations must be considered when estimating a combination of propellant materials.

There are other conclusion to be considered when developing propellants, comprising their cost, maturity, duplicability, manufacturability, compatibility, and service life. Finally, because rocket motors are commonly not used immediately after they are constructed, the propellant must keep its properties over time and beneath the environmental circumstances required for the given function.

Table 2.1: Common propellant materials and their functions

Functional Category	Common examples	Comments
Solid oxidizers	Ammonium Perchlorate (AP), other Perchlorates, Ammonium Nitrate (AN), other Nitrates, Ammonium Dinitramide (ADN), Hydrazinium Nitroformate (HNF)	Including moisture sensitivity, little ballistic, tailorability, phase transitions, and aging concerns. ADN, HNF are still immature in U.S
Energetic monopropellants	Nitramines: Cyclotrimethylenetrinitramine (RDX), Cyclotetramethylenetetramintramine (HMX), Hexanitrohexaazaisowurtzitane (Cl-20)	Nitramines are used in most class 1.1 and some 1.3 propellants. Provide increased I_{sp} These three nitramines provide similar I_{sp} in aluminized propellants. CL-20 is the denset and most oxygen-rich, but

		least mature.
Binders	Hydroxyl-Terminated Polybutadiene (HTPB), Carboxyl-Terminated Polybutadiene (CTPB), Polybutadiene Acrylonitrile (PBAN), Polyethylene Glycol (PEG), Polypropylene Glycol (PPG), Nitrocellulose (NC), Glycidyl Azide Polymer (GAP)	HTPB, CTPB, PBAN are the most common class 1.3 propellant binders. PEG, PPG, NC are generally plasticized with nitrates esters. NC and GAP are energetic
Curatives	Isocyanates, Epoxides	Isocyanates are used to cure hydroxyl terminated polymers; epoxides cure PBAN, CTPB
Fuels	Beryllium, Aluminium, Magnesium	Beryllium gives the highest I_{sp}, but is toxic. Aluminium has better I_{sp} and density than magnesium but is not as easily ignited.
Plasticizers	Dioctyl Adipate (DOA), Dioctyl Phthalate (DOP), Triacetin, other inert esters, Nitroglycerin (NG), Butanetriol Trinitrate (BTTN), TrimethyloethaneTrinitrate (TMETN), other nitrate esters	DOA, DOP, and similar esters are used as a desensitizer for nitrate ester propellants. Nitrate esters provide increased I_{sp}. NG is highest density and highest performance

		nitrate esters are used to decrease detonability (compared with NG) or to give better low-temperature properties.
Stabilizers	A02246, P-Rc-Methyl Nitroanaline (MNA), Nitrodiphenylamine (NDPA)	A02246 prevents cross linking of HTPB; MNA and NDPA stabilize nitrate esters
Ballistic modifiers	Iron Oxide, Aluminum Oxide, Oxamide	Iron oxide, aluminium oxide accelerates burn rate; oxamide and other coolants slow burn rate.

2.3 Chemical Kinetics

Ammonium perchlorate composite propellant (APCP) is the excellent solid propellant for the greater working operation of solid fuelled rocket. In ammonium per-chlorate is acompound of Ammonium per-chlorate (AP) and Powdered Aluminium (Al) and that compoundis withhold in hydroxyl terminated polybutadiene (HTPB) rubber. The Ammonium per-chlorate is act as an oxidizer whereas the powdered aluminium and HTPB rubber act as the fuel. The density of APCP is not exploded, or operated like gunpowder because it has the density three times that of gunpowder. In the chemical procedure of APCP, the primary geometry of the fuel grain and the propellant surface area transformation because the particles of the propellant are heated or vaporized on the surface due to ambient heat as far as they react and produce gas. In the combustion of APCP, the first phase is the thermal dissolution of ammonium per-chlorate: (Orr, 2009)

$$2NH_4ClO_4\ (s) + heat \rightarrow 2O_2\ (g) + N_2\ (g) + 4H_2O\ (g) + Cl_2\ (g) \tag{2.1}$$

The addition of heat leads to the aluminium fuel combusts:

$$4Al\ (s) + 3O_2\ (g) \rightarrow 2Al_2O_3 + heat \tag{2.2}$$

The reaction between aluminium and chlorine gas:

$$2Al\ (s) + 3Cl_2\ (g) \rightarrow 2AlCl_3\ (g) + heat \tag{2.3}$$

The overall reaction between oxidizer and aluminium is expressed as:

$$6NH_4ClO_4\ (s) + 10Al\ (s) \rightarrow 4Al_2O_3\ (s) + 2AlCl_3\ (g) + 3N_2\ (g) + 12H_2O\ (g) + heat \tag{2.4}$$

The thermal decomposition of HTPB can be expressed by:

$$HO(C_4H_4)_{50} + heat \rightarrow 49C_4H_4\ (g) + 2CO(g) + C_2H_6\ (g) \tag{2.5}$$

Combustion of HTPB decomposition products:

$$C_4H_4\ (g) + 5O_2\ (g) \rightarrow 4CO_2\ (g) + 2H_2O + heat \tag{2.6}$$

Solid fuelled rocket has a very complicated chemical kinetics procedure. The burning of good chemical kineticsprocedure can be formulated as in the terms of linear burn rate $(r, m/s)$, the pressure exponent $(n, m/s)$, and the non-dimensional combustion pressure $(\bar{P})$, and the pressure coefficient (k):

$$r = k\bar{P}n \tag{2.7}$$

The mass flux through the nozzle is expressed in the terms of surface area (A, m^2) and the average density of the APCP $(\rho, Kg/m^3)$:

$$m = \rho Ar \tag{2.8}$$

3. SOLID FUELLED ROCKET ENGINE

Good Rule For Rocket Experimenters To Follow Is This: Always Assume That It Will Explode.

A jet engine that performs on solid fuel is a totally self-possessed device that converts chemical energy into kinetic energy in a disciplined way. In a solid-propellant rocket engine, all the propellant is situated in the combustion chamber as a single charge. The engine is commonly operated all the time until the propellant is completely gone. Thermodynamically a solid fuelled rocket motor is alike to a liquid fuelled engine. Solid propellant rocket engines were the first rocket engines to acquire practical utilization. Solid propellant rocket engines are nowadays used as primarily in missiles, commercial satellites and in spacecraft (as rocket engines for the first stage in launch vehicles). Solid propellant rocket engines are classified by high reliability (99.6%-

99.9%), a long storage life in launching condition, the accessibility of considerable thrust for a very short combustion time, safety in controlling and a high propellant density (1.5-2 g/cm^3).

Figure 3.1: Solid fuelled rocket engine

3.1 Basic Configuration

An elementary solid propellant rocket engine consist of housing, nozzle, grain (propellant charge), and igniters. The housing is normally made of steel, but the fibreglass may also be used. The grains respond like a solid mass, burning in a expected manner and generating exhaust gases. The nozzle dimensions are computed to keep a design chamber pressure, whereas producing thrust from the exhaust gases. The section of the jet nozzle that experiences the greatest thermal loads is made of graphite, refractory metals, or alloys of refractory metals. The super critical section of nozzle is made of graphite, plastic, or steel. New designs may also cotain a steerable nozzle for direction, avionics, rescue hardware (parachutes), self-destruct mechanism, controllable deviation and attitude control motors, and thermal management materials. The elementaryeasiness of the solid propellant rocket make possible wide application. In accession, the dependability is very high, due to small number of exclusive components balanced with a liquid-fuelled rocket engine. The exhaust velocity of that type of engines is not very high (about 2,700m/s), but the lack of turbo-pumps and independent fuel tanks, and the absence of complex valves and pipelines, can generate high mass ratio and low cost or both. (Turner, 2009)

3.2 Design of solid propellant rocket engine

A solid propellant rocket engine is formed by five main components:

- A case containing the solid propellant and opposing internal pressure when the rocket is operating.
- The solid propellant charge (or grain), which is commonly bonded to the inner wall of the case, occupies the greater part of its volume before ignition.
- At the time of burning, the solid propellant is converted into hot combustion output. The volume occupied by combustion products is called combustion chamber.
- The nozzle channels: the expulsion of the combustion output and because of its shape; it accelerates the combustion output to supersonic velocity.
- The igniter, which can be a small rocket, starts the rocket operating when an electric signal is received.

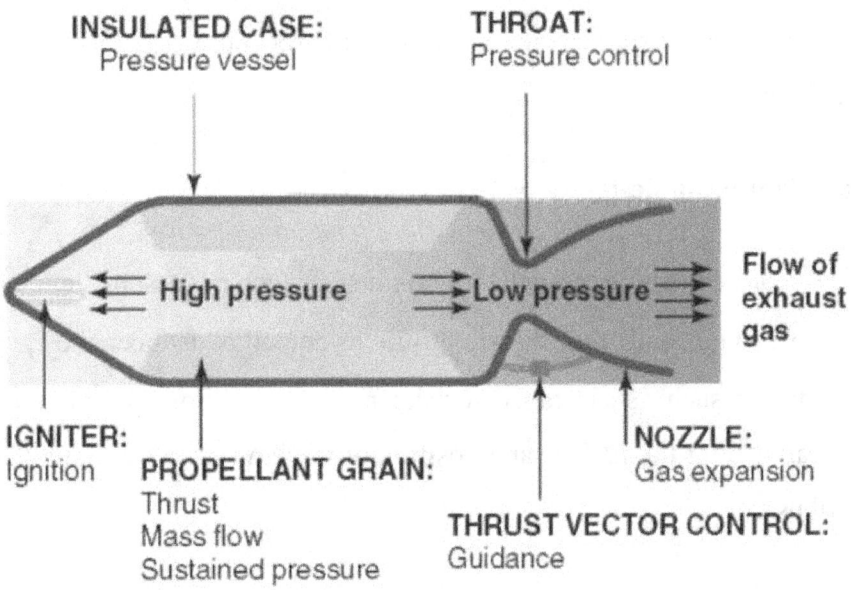

Figure 3.2: Solid fuelled rocket engine design

3.2.1 Solid fuel rocket engine case

The solid rocket motor case is the pressure holder that includes the solid propellant and maintains a structural port to external components. The case is designed to hold the high pressure generated by the burning propellant throughout motor operation. It is also the airframe that transmit the thrust from the motor to the launch vehicle or missile system.

Metal or composite materials are commonly used in case design. Metals are commonly less costly, more damage liberal, and better characterized the composites. Composites materials normally weigh less due to high strength to weight ratio but are less constrained for corresponding thickness. A composite case can be as much as five time lighter than metal case. This weight deduction requires less propellant for comparable vehicle performance, which eventually can lead to less costly motor. Solid fuel rocket cases are of two types: filament-wound composites and metal.

- For filament-wound composites cases, the insulation is located over a removable mandrel, cured, and machined, and the filament-resin combination is twisted over it. After the case is charged, the mandrel is removed.
- In a metal case motor, the case is constructed first, and the insulation is set up inside the case and then cured. This curing procedure also connects the insulator in the case.

3.2.2 Propellant grain design

The essetial requirements for propellant grain design such as reasonable length, volume, maximum pressure, and the thrust profile principally deduce the grain design. Additional factor such as clearance for the nozzle, the thrust vector control (TVC) method, clearance for the igniter, and position of the ignition system are also important considerations.

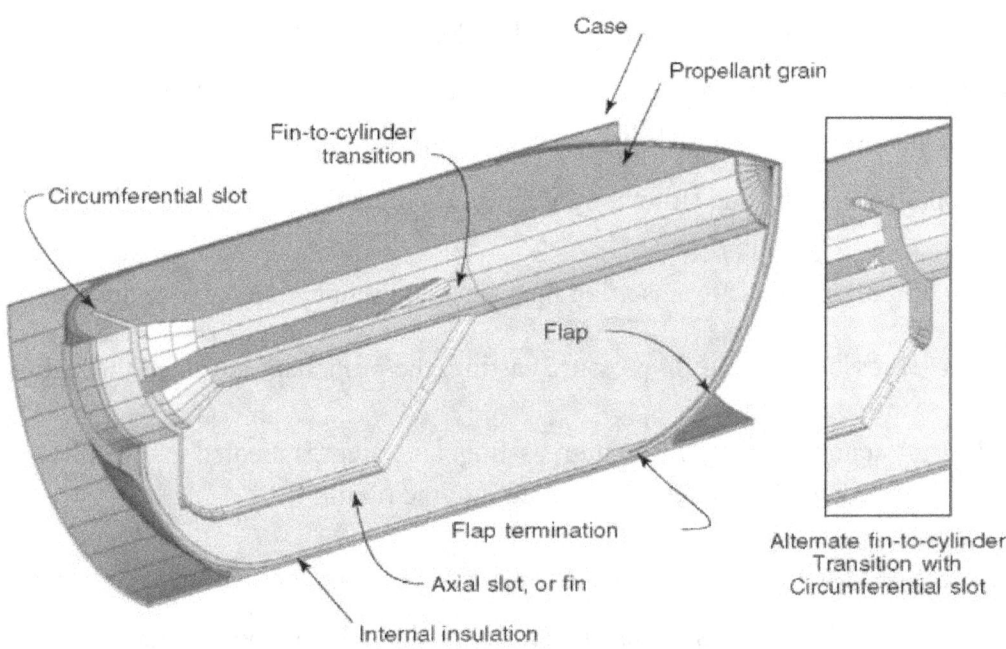

Figure 3.3: Features used in propellant grain design

Figure (3.3) shows an example of some features used in grain design. Grain structural loads many times require stress relief characteristics such as flaps or slots. Feature such as radial or longitudinal slots are structuredinto the propellant grain to gain the desired motor operation. The structural need of the propellant grain design are reasonable grain thickness, accessible distance, and the presence or absence of flaps and stress relief slots. A flap is an elastomeric piece attached to the grain that detach the grain form the case. Flaps characteristically do not have a large effect on the ballistic behaviour of the motor; though, stress relief slots can face considerable design complications. Stress relief slots are slots situate in the grain for structural purpose.

Structural investigation of the solid propellant grain is essential because structural breakdown of the propellant can result a sudden disaster of the rocket motor. A crack in the propellant or a detachment in the propellant-to-case bond will increase the surface area, provoking increased pressure, which may blow up the case. Another disaster mode is casing seal design. Seals are involve in casings that have to be accessible to load the grain. Once a seal weaken, hot gas will corrode the escape path and results in failure. This was the cause of the Space Shuttle Challenger disaster. In designing a motor, the stresses and strains actuated in the propellant by these loads must be kept beneath the propellant's potentiality. Most methods for decreaseing grain stresses and strains decrease the quantity of the propellant in the motor, so preciseattention must be given to both propellant loading and structural integrity. (Wikipedia)

The two primary methods for inserting the propellant into the motor case are

(1.) Pressure Casting: In pressure casting, the propellant is forced through a tube into the motor.
(2.) Vacuum Casting: In vacuum casting, a vacuum is created inside of the motor case, and the propellant is pulled into the motor.

Another type of grain fabrication is extrusion. The propellant is forced over a die and cut to length. This process is not generally used for large rocket motors due to the complication of holdingthe grain in the rocket motor case. Casting tooling may be detachable; the tooling is detached from the motor after propellant insert is accomplish or may be left in place. Sometimes both type of grain developing are used in the same motor.

Structural requirement may require curing the propellant under pressure. The combination of vacuum casting and pressure cure results in grains that have few fault such as cavities. Grain may be cast with simple tooling, and more complicated slots are machined into the grain after propellant cure. This results in an additional manufacturing step, but changing the grain is as simple as altering the machine program, somewhat altering expensive hard tooling.

3.2.3 Propellant Grain Profile

Solid-fuelled rocket burns with extreme heat and intense light from the surface of exhibited propellant in the combustion chamber. In this fashion, the profile of the propellant inside the rocket motor plays is an important charactersticsin the overall motor operation. As the surface of the propellant burns, the shape emerges, most often adapting the propellant surface area disclosing into the combustion gases. The mass flux

(Kg/sec.) of combustion gases induced is a function of the immediate surface area A_s *(m²)*, and linear burn rate b_r *(m/sec.)*:

$$\dot{m} = \rho . A_s . b_r \tag{3.1}$$

The thrust profiles related with the shapes in figure (3.4) can be understood from simple shape arguments.

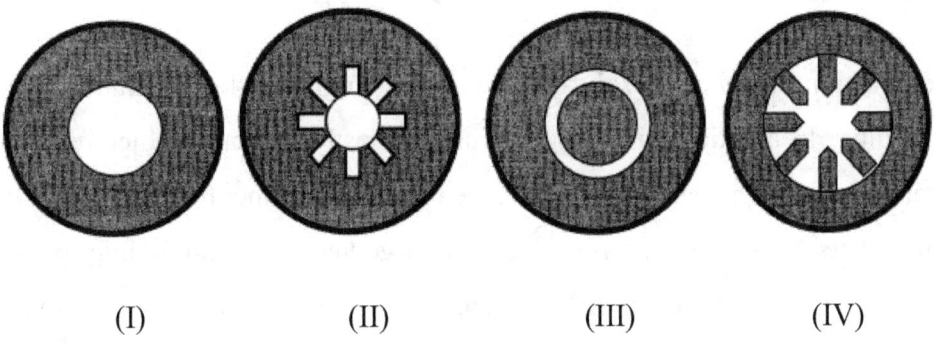

(I)   (II)   (III)   (IV)

Figure 3.4: Cross sections of propellant grain

Type (I) is called 'progressive' is the simplest propellant grain shape to understand. The circumference of the circular cross-section expands linearly with time, as does the area of the burning surface, and there is linear increment in mass flow rate and therefore in thrust.

Type (II) propellant grain shape is most generally used in solid-fuelled rockets. It creats a quasi-constant thrust, because the primary burning area is very large due to the binding of the cog shape; as the cog 'teeth' burns, the loss of burning area is balanced by the increasing area of the cylindrical part. This type of propellant grain profile is simple to mold, and is very energetic in producing an almost constant mass flow rate.

Type (III) propellant grain profile develops a completely flat thrust profile, because burning take place both on the exterior surface of the inner rod and on the inner surface of the outer cylinder. The decrement in burning area of the exterior surface of the outer rod is exactly corrected by the increasing burning area on the inner surface of the cylinder. This type of propellant grain profile is difficult to manufacture.

Type (IV) propellant grain profile is an incrediblegrain profile. It can be used to adjust the thrust profile for a certainpurpose. The narrow fins of propellant primarily generate a very high surface area, and so the thrust at the begning is very high. Once they have burned away then a low and slowly devloping thrust is generate by the cylindrical section. Such a profile may be useful for powerful acceleration followed by sustained flight. (Turner, 2009)

3.2.4 Combustion chamber

The combustion chamber of solid-fuelled rocket engine is larger than liquid-fuelled rocket engine. In addition, since high thrust is mostly the main objective, the throat diameter is larger. The pressure experienced by each of them is almost the same in modern rockets-50 bar. However, designing a large vessel to contain high pressure and high temperature is much more difficult than designing a smaller vessel. As in the case of solid-fuelled combustion chamber, the temperature of combustion is much higher than the salving point of most metals. The combustion elements cannot be permotted to contact the walls for any expanded period, or disaster will result. The best method is to connect the propellant to the walls and to cover-up the unexpended inside surfaces with a intractable insulating layer. This process is known as case bonding, and is used in most latest solid motors. The grains burns only on its innner surface, so the propellant act as an insulator.

3.2.5 Insulation

Internal insulation is the heat obstruction between the case and the propellant. It is insulate the case from reaching the temperatures that would expose its structural integrity.

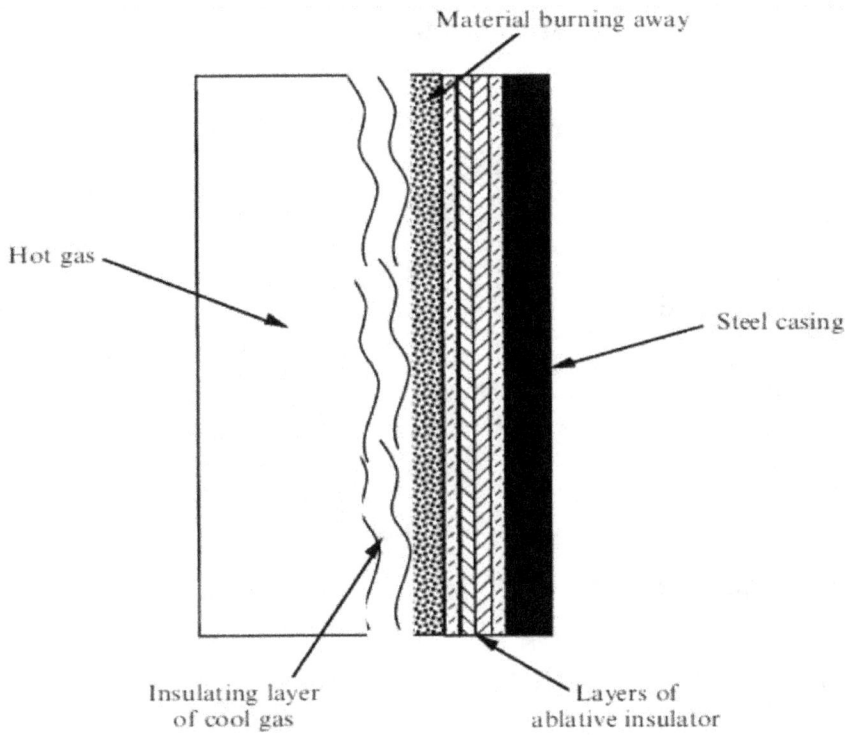

Figure 3.5: Insulating layers for thermal protection

Insulation also serves to

(1.) Reserves the circulation of case stresses to the propellant.

(2.) Block burning on selected propellant surfaces.

(3.) Accommodate a pressure seal for the case.

(4.) Control the diffusion of chemical components to or from the propellant.

Insulation may also be situated on the exterior of the rocket motor to protect the case from aero-heating and to contribute damage consideration and protection from the elements. Insulator materials are organic elements that consist of reinforcing fillers stored in a binder. Fillers provide to char-layer durability and include silica, asbestos, and carbon or Kevlar fibres, nylon and glass cloth. Elastomers and plastics are two classes of binders.

Internal insulator design especially involves arrangement for material char and ablation. The general equation for computing design thickness is (Turner, 2009)

$$Thickness = (ET \times MAR) \times SF + TP + MT \tag{3.2}$$

Where

ET = Exposure time

MAR = Material affected or char rate

SF = Safety factor

TP = Thermal protection thickness

MT = Manufacturing tolerance

3.2.6 Igniter

The reason of igniter is to supply the heat and pressure quickly needed to start propellant combustion. Figure (3.6) shows a illustrative solid propellant rocket motor that has an axial flow igniter and some of the factors that affect motor ignition. These factors implicate igniter position, mass flow rate and action time of igniter, an effect of igniter output on the propellant surface, propellant grain geometry, combustion chamber free volume, and nozzle throat size.

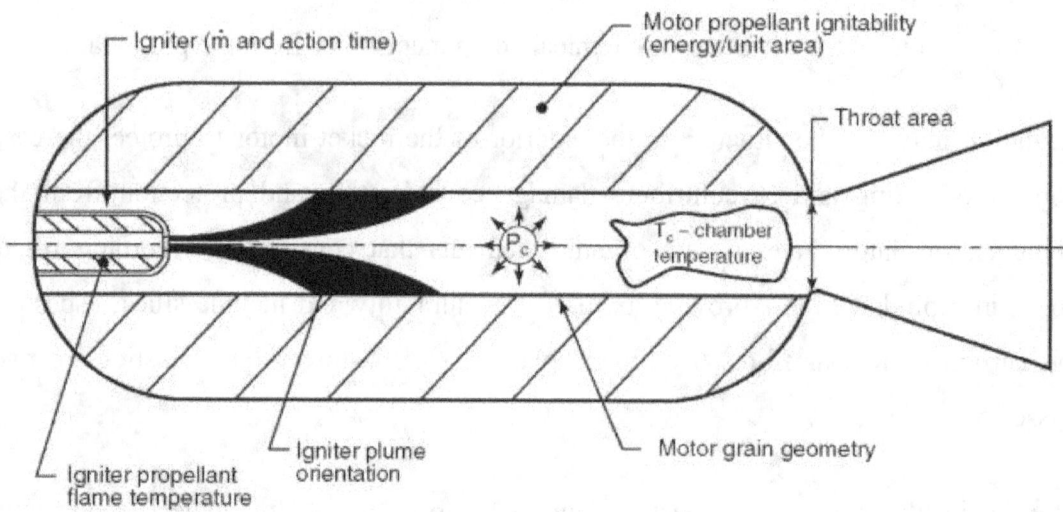

Figure 3.6: Ignition and grain heating of a solid propellant rocket motor

There are four main components in an ignition system:

- The safety device
- The electro-explosive device
- The booster charge
- The igniter main propellant charge

The safety device contribute the electrical and mechanical safety characterstics needed to hold on electrical current idly from causing accidental ignition. For planned ignition, electrical arming signals are sent to the safety device that mechanically and electrically arm or align it. In this arm situation, an electrical firing signal can start a chain of processes that cossumates in motor ignition. The electrical firing signal is sent to an electro-explosive mechanism called an initiator. The output from the initiator generally ignites a booster charge. The output from the booster charge then ignites the igniter main propellant charge that in turn ignites the solid propellant in the rocket motor.

3.2.7 Nozzle

A rocket engine nozzle is a propelling nozzle usually of the de Laval type used in a rocket engine as shown in figure (3.7) to expand and accelerate the combustion gases, from burning propellants, so that exhaust gases exit the nozzle at hypersonic velocities. The function of nozzle is to provide a ballistic throat for the motor, control motor pressure, direct subsonic gases into the throat, additional thrust, and thrust vector control (TVC). A convergent-divergent nozzle design accelerates the exhaust gas out of the nozzle to produce thrust. The nozzle environment is harsh; temperatures can reach $6000°F$. Pressures can exceed 3500 psi, exposure time can reach 300 sec.

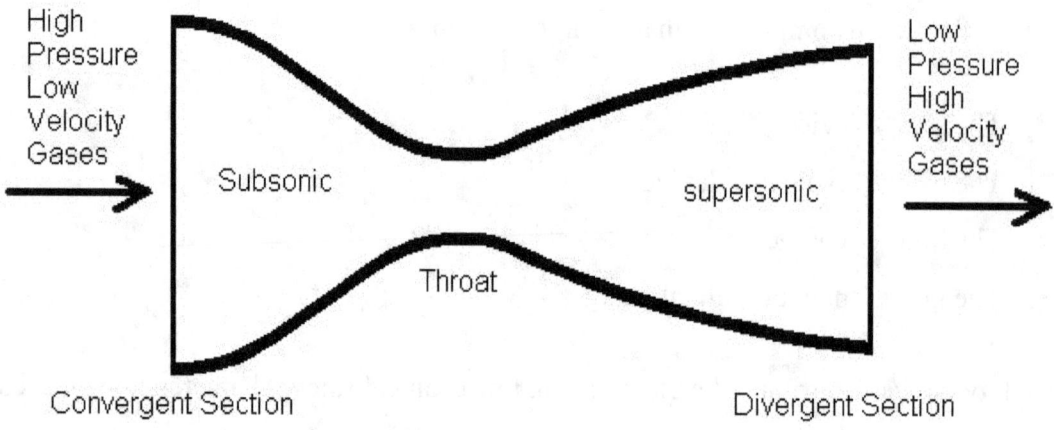

Figure 3.7: A convergent-divergent nozzle (de-Laval Nozzle)

In convergent section of de-Laval Nozzle a high pressure low velocity gas is enter and flows at subsonic speed (M < 1). At this section, the speed of the subsonic flow of gas will increase and mass flow rate is constant. At the throat of the nozzle, the cross sectional area of the nozzle is minimum and at this stage the velocity of the gas becomes sonic (M=) and the flow of the gas is called choked flow. After this the cross sectional area of the nozzle increases and the gas begins to expand and the gas flow becomes supersonic 'low pressure high velocity gas' (M > 1). The gas flow through the entire de Laval Nozzle is isentropic.

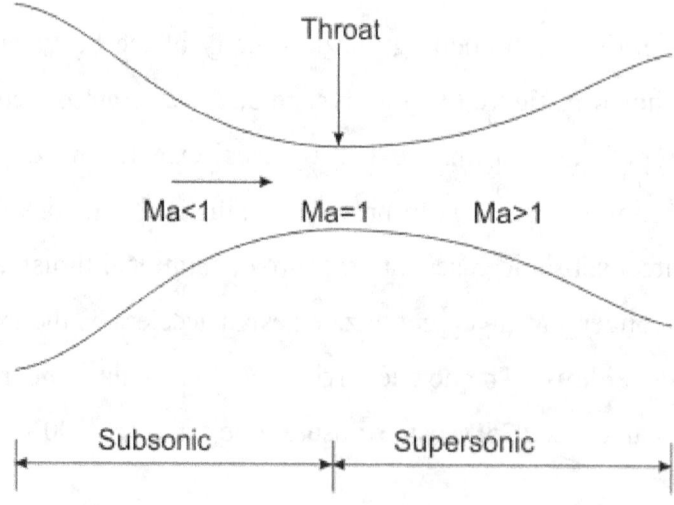

Figure 3.8: Flow process inside the nozzle

The nozzle must be constructed from a material that can withstand the heat of the combustion flow. Temperatures, pressures, and gas velocities greatly vary from the inlet region of the nozzle through the throat and into exit cone. Often, heat-resistant carbon based materials are used, such as amorphous graphite or carbon. (Sutton, 2001)

3.3 Thrust Vector Control System

Thrust vector control (TVC) uses exterior means (normally mechanical or fluidic) to accommodate the direction of the thrust, although changing the direction of the rocket's flight path. Figure (3.7) shows how one type of TVC, a movable nozzle, set up a turning moment on the rocket. TVC system can be designed to provide control in all three axes: pitch (up and down motion), yaw (side to side motion), and roll (rotation about the longitudinal axis of the rocket).

Fixed nozzle mechanism use thrusters, secondary fluid injection, or mechanical deflectors to adjust the thrust vector direction. Thrusters are multiple dictinct nozzles placed in the criticalpositions on the missile to bring out the desired turning forces. The gas may be delivered by a pressurized bottle (cold gas), a gas generator (warm gas), or may be occupiedstraightly from the combustion chamber of the rocket motor (hot gas). Thrusters have very fast reaction time, and they also generate a shock wave outside of the rocket that can magnify the resulting force. Disadvantages embrace the valuableextrinsic materials essential for hot gas valves and the weight and volume disadvantage of wrapping.

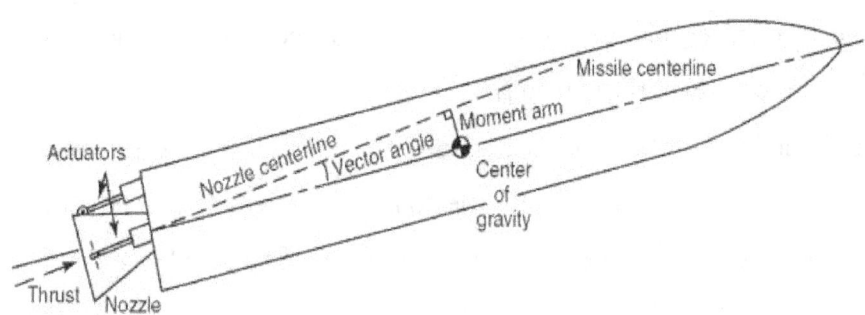

Figure 3.9: Thrust vector control system mechanism

3.3.1 Secondary Fluid Injection

In secondary fluid injection TVC, a fluid (liquid or gas) is inserted into the exit cone over the wall. This constitute side forces from a combination of the thrust of the injecting jet and pressure instabilities from shock waves. The injectant can be an inert or excitable liquid, in which a chemical reaction effects in further side force. Gas injectants may come from gas generators (warm gas), or they may be bled straightly from the motor combustion chamber (hot gas). Advantages involve fast feedback and the thrust extension to the main flow. Disadvantages are the large occupying volume required and finite thrust deflection (about $6°$).

3.3.2 Mechanical Deflectors

TVC systems that use mechanical deflection include jet vanes, jet tabs, and jet evators.

- Jet vanes are aerodynamic fins internally the nozzle that revolve around to contribute pitch, yaw, and roll control. Roll control is one advantage of jet vanes. Additional advantage is the low torque needed to drive jet vanes. Disadvantages involve the thrust decrement due to the aerodynamic drag on the vanes.
- Jet tabs are renounceable surfaces boosted at the aft end of the exit cone that rotatein and out of the gas flow, generating shock waves and pressure instability. An advantage across jet vanes is that no thrust losses appear when the tabs are exposed of the flow. Jet tabs also have anignoble actuating force condition. In spite of this, jet tabs do not produce any roll control, have large thrust losses, and also require extrinsic materials.
- The jet avator consists of a spherical ring mounted around the nozzle exit cone that can be rotated into the supersonic gas stream. This rotation creates a steering side force on the rocket. One advantage is a side force that is linear with the deflection angle. Disadvantages include large weight and volume, severe environment requiring exotic materials, and large thrust losses.

3.3.3 Movable-Nozzle TVC

In movable-nozzle systems, the nozzle is automaticallyrotated, which deviate the hot supersonic flow of gases, therefore changing the thrust vector. Portable nozzles have lesser thrust losses than the other types of TVC. However, a single movable nozzle cannot maintain any roll control. Roll needs at least two nozzles. Movable nozzles are also subcategorized compatible to the location of the joint.

3.3.4 Flexible Bearing or Flexbearing

The flexible joint (also known as a flexible bearing or flexbearing) normally consists of interchangingcoating of an elastomeric material for elasticity and a inflexible material (steel or composite) for strength and hardness. The expected and repeatable character of the driving force needed to actuate a flexible bearing nozzle is treated as an advantage. Flexbearing nozzles are the most extensively used TVC systems.

3.3.5 Ball-and-Socket Joint

The ball-and-socket joint is also noted as a trapped ball, has a spherical socket that mounts inside a coupling spherical ball surface. The ball-and-socket nozzle has reliable advantages throughout the flexbearing, containing higher vector angles, higher motor pressure efficiency, and less axis point shift. Disadvantages involve an unstable stick-slip friction force and it directs an anti-rotational device to avoid the nozzle from rolling in the socket.

3.3.6 Fluid Bearing/Rolling Seal

It is also known by the patented name Techroll, is possessed of a pair of rolling elastomeric convolutes that include a fluid. The greatest advantage is the low drivingforce needed to move the bearing. The main disadvantage is the low structural hardness and resulting large misarrangement of the nozzle.

3.3.7 Hinged Movable Nozzle

The hinged movable nozzle is supported on the thrust pins that ride in journal bearings. The hinged nozzle has the advantage of the high deflection capability and low actuating force. The main disadvantage is that it provides control in only one direction (pitch or yaw) because it rotates only about one axis.

3.3.8 Gimbaled or Hinged Nozzle

The gimbaled nozzle is an expansion of the hinged nozzle; the thrust pins about which the nozzle rotates are themselves installed in a rotating assembly that rotates about different set of thrust pins which are situated at 90°just about the nozzle from the first set. This gives the nozzle Omni-axial motion ability. The gimballed nozzle has been experimented in the ground and in flight but has not been in manufacturing.

3.3.9 Rotatable Nozzle

The rotatable nozzle is a tipped nozzle located on a rolling bearing so that it can rotate around the motor centreline. Its main advantage is the low driving force needed. However, it is bounded to motors that have various nozzles because movement of the nozzle creats pitch, yaw, and roll moments that must be stabled by the other nozzle. Another disadvantage is that the rotationaldeviation required is much better than the absolute vector angle attained. (Sutton, 2001)

Table 3.1: Thrust vector control mechanism

Type	Advantages	Disadvantages
Liquid-side injection	Proven technology; specific impulse of injectant nearly offsets weight penalty; can check out before flight; easy to adapt to various motors; duration limited by liquid supply; ±6°; components are reusable	Toxics liquid are needed for high performance; potential spills and toxic fumes with some propellants; difficult packaging for tanks and feed system; sometimes requires excessive maintenance; limited to low vector angle applications
Hot-side-gas	Light weight; low performance	Technology is not yet

injection	loss; low actuation power; low compact/volume	proven; multiple hot sliding contacts and seals in hot gas valve; limited duration; hot piping expansion; requires special hot gas valves
Jet vanes	Proven technology; roll control with single nozzle; low actuation power; ±9°	Extends missile length; limited duration; thrust loss of 0.5 to 3%; erosion of jet vanes
Jet tabs	Proven technology; compact package; low actuation power	Thrust loss, but only when tab is in the jet; erosion of tabs; limited duration
Jetavator	Proven on polarise missiles; low actuating power; can be lightweight	Limited duration; thrust loss and erosion; hot gas circulation
Movable nozzle (flexible bearing)	Proven technology; predictable actuating power; up to ±12°; no sliding, moving seals	High torque at low temperatures; high actuation forces; variable actuation force
Movable nozzle (rotary ball with gas seal)	Proven technology; no thrust loss if entire nozzle is moved; ±20° possible	Limited duration; highly variable actuation power; needs continuous load to maintain seal; sliding, moving hot gas spherical seal
Gimbal or Hinge	Simple proven technology; very small thrust loss; low torques; duration limited only by propellant supply; low power; ±12°	Requires flexible piping; high inertia; large actuators

4. THERMODYNAMICS OF SOLID ROCKET PROPULSION

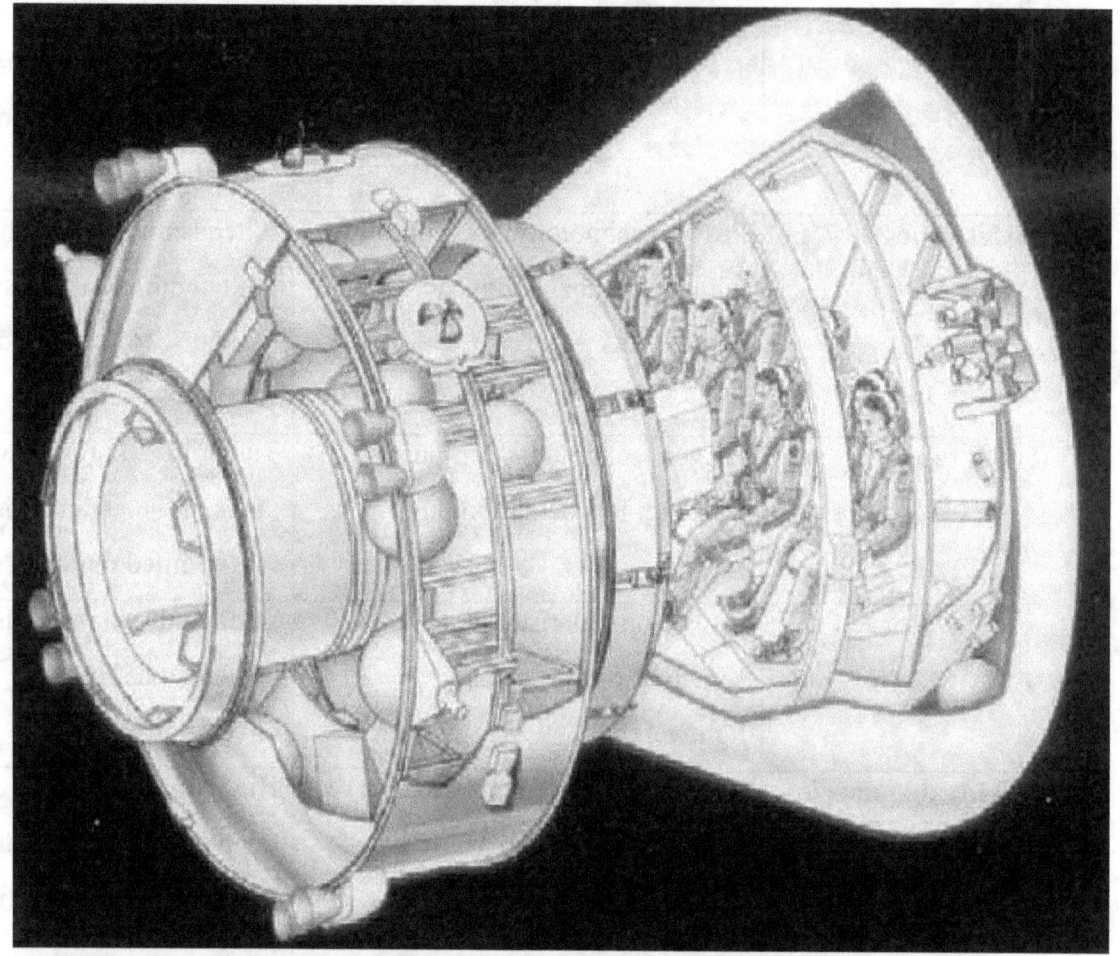

It is difficult to say what is impossible, for the dream of yesterday is the hope of today and reality of tomorrow.

In this chapter we introduce the concepts and laws of thermodynamics and quasi-one-dimensional flow. For the rocket propulsion, it is very important to understand the thermodynamic relations and the study of substances in equilibrium, including thermal, mechanical, and chemical equilibrium. The limitation of equilibrium totally depends upon the flow processes, and the substances involved in flow processes are not in equilibrium. The losses related to the performance of flow processes can be identified with lack of equilibrium and the quantitative prediction of such losses is beyond the scope of thermodynamics. For solid rocket propulsion the measured idle value of performance of the rocket is usually between 1 to 6%. (Oates, 1997)

The thermodynamic principles can described as simple mathematical relationships, as in the case of the idle rocket propulsion systems. In designing a new rocket, it is very necessary and useful to use the idle rocket parameters. These are the following assumptions to design an idle rocket unit:

- All the chemical reaction products are homogeneous.
- All the chemical reaction products obey the perfect gas law.
- The flow is adiabatic (no heat transfer across the rocket wall).
- All the boundary layer effects are neglected and no considerable friction.
- In the nozzle flow, there is no shock waves and discontinues in the nozzle flow.
- The flow of propellant is steady and constant.
- The expansion of the working fluid is uniform and steady.
- The start-up and shut-down are of very short duration.
- All exhaust gasses leaving the rocket have an axially directed velocity.
- The gas temperature, velocity, pressure, and density are uniform across any section normal to the nozzle axis.
- The propellants are stored at room temperature.
- Chemical equilibrium is established within the rocket chamber.

4.1 Laws of Thermodynamics

Thermodynamics mainly concerned withheatand work interactions. In his section we discuss the laws of thermodynamics. These are the following laws of the thermodynamics:

4.1.1 The zeroth-law of thermodynamics

Thezerothlaw of thermodynamicsstates: **"Iftwo bodies are separately in thermal equilibrium with third body, they are in thermal equilibrium with each other"**. If we brought a hot body and put it into contact with a cold body, the transfer of the temperature from the hot body to cold body take place until the hot body stops getting colder and the cold body stops getting hotter. At this stage both bodies are in thermal equilibrium.

4.1.2 The first law of thermodynamics

The first law of thermodynamics is also known as the law of conservation of energy. It states: **"Energy can be transformed from the one form to another such as work, heat etc. it cannot be created or destroyed"**. The total amount of energy and matter in the universe remains constant. In any thermodynamic process, increment in the internal energy of a system is equal to the difference between increment of heat of the system accumulated by the system and increment of work done by it. If we assume that dQ is the increment of heat and dW is the increment of work done by it, the first law of thermodynamics can written as

$$dU = dQ - dW \tag{4.1}$$

where dU is the change or increment in the internal energy of the system.

4.1.3 The second law of thermodynamics

The second law of thermodynamics is a law of entropy. It states: "The mechanical work can be derived from a body only when the body interacts with another body at lower temperature and any process occurring without external cause results in an increase of entropy". The second law of thermodynamics governs the relations between the stages of energy in a closed system. Thus,

$$S_2 - S_1 = \int_1^2 \frac{\delta q}{T} \tag{4.2}$$

or

In differential form

$$TdS = \delta q \tag{4.3}$$

Inside a rocket nozzle and combustion chamber thermodynamic relations needed to calculate the performance and determine parameters of rocket propulsion system. The use of these thermodynamic relations should give a basic understanding of thermodynamic process involved in rocket gas behaviour and expansion. Thermodynamic entropy is defined as (in differential form):

$$dS \equiv \frac{\delta q}{T} \tag{4.4}$$

Because flow through a nozzle is a reversible process and it is an isentropic process. The in internal energy (E), pressure (P), enthalpy (H), and volume (V) related to the heat flux by the law of energy conservation. This can be represented in differential form:

$$dQ = dE + pdV = dQ = dH - Vdp = 0 \tag{4.5}$$

Because the change in entropy is zero, the heat transfer must be zero. This leads to the following thermodynamic relationships:

$$dE = C_v dt \tag{4.6}$$

$$dH = C_p dt \tag{4.7}$$

where C_v and C_p are specific heat coefficient at constant volume and at constant pressure respectively. Substituting the eqns. (4.6) and (4.7) into eqns. (4.5) and dividing by temperature (T):

$$dS = \frac{dQ}{T} \tag{4.8}$$

$$dS = C_v \frac{dT}{T} + R\frac{dV}{V} = C_p \frac{dT}{T} - R\frac{dP}{P} = 0 \tag{4.9}$$

Equation (4.9) can be integrated to yield an explicit solution evaluated between point 1 and 2:

$$S_2 - S_1 = C_v \ln\frac{T_2}{T_1} + R \ln\frac{V_2}{V_1} = C_p \ln\frac{T_2}{T_1} - R \ln\frac{P_2}{P_1} = 0 \quad (4.10)$$

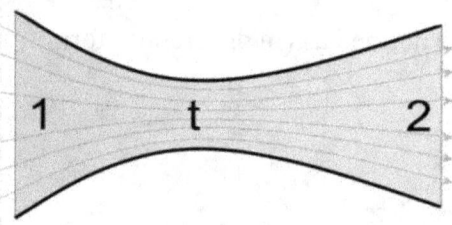

Figure 4.1: Flow through a de Laval Nozzle

The thermodynamic heat ratio (γ) and gas constants (R) are related to C_p and C_v:

$$\gamma = \frac{C_p}{C_v} \quad (4.11)$$

$$R = C_p - C_v \quad (4.12)$$

$$\bar{R} = \frac{R}{M} = \frac{C_p - C_v}{M} \quad (4.13)$$

For rocket motor combustion products, $\gamma = 1.20$. On solving the equation (4.10), we find temperature dependence pressure:

$$\ln\frac{T_2}{T_1} = \frac{R}{C_p} \ln\frac{P_2}{P_1} \quad (4.14)$$

$$\frac{T_2}{T_1} = \left(\frac{P_2}{P_1}\right)^{\frac{R}{C_p}} = \left(\frac{P_2}{P_1}\right)^{\frac{\gamma-1}{\gamma}} \quad (4.15)$$

The local sonic velocity (a, m/s) can be expressed as:

$$a = \sqrt{\gamma \bar{R} T} \tag{4.16}$$

Where $\bar{R}$ is the specific gas constant for the local gas and has unit *J/Kg-C*. The Mach number (M) at a point describes a velocity relative to its local sonic speed. The Mach number is defined as:

$$M \equiv \frac{v}{a} = \frac{v}{\sqrt{\gamma \bar{R} T}} \tag{4.17}$$

If we want to examine the energy of two distinct points in a flow, the first point has a very low velocity such that the great majority of its energy is in the form of heat. The corresponding stagnation temperature T_o and specific heat capacity C_p must equal the energy present at some point x in the form of thermal and kinetic energy. Therefore, if flow losses are negligible, energy continuity must apply:

$$C_p T_o = C_p T_x + \frac{V_x^2}{2} \tag{4.18}$$

On dividing the above eqn. by $C_p T_x$, we get:

$$\frac{T_o}{T_x} = 1 + \frac{V_x^2}{2 C_p T_x} = 1 + \frac{\gamma-1}{2} \frac{V_x^2}{\gamma \bar{R} T} = 1 + \frac{\gamma-1}{2} \frac{V_x^2}{a_x^2} \tag{4.19}$$

$$\frac{T_o}{T_x} = 1 + \frac{\gamma-1}{2} M_x^2 \tag{4.20}$$

This thermodynamic relationship can be used to find pressure dependence on Mach number between two distinct points 1 and 2. In the figure (4.1), point 1 is found as the stagnation point in the combustion chamber of rocket motor and point 2 can be any point "downstream". According to the equation (4.20), the temperature ratio at the point 1 and 2 can be calculated: (Oates, 1997)

$$\frac{T_1}{T_2} = 1 + \frac{\gamma-1}{2} M_2^2 \tag{4.21}$$

Rearranging the eqn. (4.15) and solving for pressure and substitute the value of eqn. (4.21), the relationship between the exit Mach number and the pressure ratio can be computed:

$$\frac{P_1}{P_2} = \left(\frac{T_1}{T_2}\right)^{\frac{\gamma}{\gamma-1}} \tag{4.22}$$

$$\frac{P_1}{P_2} = \left(1 + \frac{\gamma-1}{2} M_2^2\right)^{\frac{\gamma}{\gamma-1}} \tag{4.23}$$

On solving for Mach number, we get:

$$M_2 = \sqrt{\frac{2}{\gamma-1}\left[\left(\frac{P_1}{P_2}\right)^{\frac{\gamma-1}{\gamma}} - 1\right]} \tag{4.24}$$

The exhaust velocity can be easily calculated in terms of the exit Mach number:

$$v_e = v_2 = M_2\sqrt{\gamma \bar{R} T_2} \tag{4.25}$$

4.2 Nozzle Flow Equation

The development of nozzle flow equation is basically depending upon the some thermodynamic relationships. Let us consider a channel of varying area duct and a calorically perfect gas is flowing through it. If the mass flow rate is $\dot{m}$, it can be obtained from the continuity equation in terms of Mach number, local area, and stagnation properties:

$$\dot{m} = \rho_t u_t A = \rho_t a_t A \frac{\rho}{\rho_t} \frac{a}{a_t} \frac{u}{a} \tag{4.26}$$

$$\frac{\rho_2}{\rho_1} = \frac{P_2}{P_1} \frac{T_1}{T_2} \tag{4.27}$$

$$\frac{\rho_2}{\rho_1} = \left(1 + \frac{\gamma-1}{2}M^2\right)^{\frac{1}{\gamma-1}} \tag{4.28}$$

$$\dot{m} = \frac{AP_t}{\sqrt{T_t}}\sqrt{\frac{\gamma}{R}}\left(1 + \frac{\gamma-1}{2}M^2\right)^{-\left[\frac{(\gamma+1)}{2(\gamma-1)}\right]} M \tag{4.29}$$

In terms of local static pressure the above equation (4.30) is expressed as:

$$\dot{m} = \frac{AP_t}{\sqrt{T_t}}\left(\frac{P}{P_t}\right)^{\frac{1}{\gamma}}\left[\frac{2}{R}\frac{\gamma}{\gamma-1}\left\{1 - \left(\frac{P}{P_t}\right)^{(\gamma-1)/\gamma}\right\}\right]^{\frac{1}{2}} \tag{4.30}$$

$$\dot{m} = \frac{A^* P_t^*}{C^*} \tag{4.31}$$

Where * denotes the unity of Mach number (M = 1) at the throat of the nozzle and C^* is the characteristic velocity.

$$C^* = \frac{P_t A}{\dot{m}} = \frac{\sqrt{\gamma R T_t}}{\gamma\sqrt{[2/(\gamma+1)]^{(\gamma+1)/(\gamma-1)}}} \tag{4.32}$$

On dividing the equation (4.30) and (4.31) by the equation (4.32) respectively, we got the expression for the area variation with Mach number and static pressure:

$$\frac{A}{A^*} = \frac{P_t^*}{P_t}\sqrt{\frac{T_t}{T_t^*}}\left[\frac{2}{\gamma+1}\left(1 + \frac{\gamma-1}{2}M^2\right)\right]^{(\gamma+1)/2(\gamma-1)} \frac{1}{M} \tag{4.33}$$

$$\frac{A}{A^*} = \sqrt{\frac{\gamma-1}{2}}\left(\frac{2}{\gamma+1}\right)^{(\gamma+1)/2(\gamma-1)} \frac{P_t^*}{P_t}\sqrt{\frac{T_t}{T_t^*}}\left(\frac{P_t}{P}\right)^{1/\gamma}\left[1 - \left(\frac{P}{P_t}\right)^{\frac{(\gamma-1)}{\gamma}}\right]^{\frac{1}{2}} \tag{4.34}$$

4.3 Rocket Nozzle Performance

The variation in the thrust levels inside the rocket nozzle is mainly caused by the changing ambient pressure with altitude. As we know that inside the rocket nozzle, the flow is isentropic and the gas is calorically perfect. These approximations can be utilized to obtain the estimate of such variations. The relationship between the first law of thermodynamics and isentropic process is given by:

$$\frac{v_e^2}{2} = h_c - h_e = \frac{\gamma}{\gamma-1} RT_c \left[1 - \left(\frac{P_e}{P_c}\right)^{\frac{(\gamma-1)}{\gamma}}\right] \quad (4.35)$$

Subscript 'c' shows the stagnation condition within the rocket chamber.
Hence

$$v_e = \sqrt{\frac{2\gamma}{\gamma-1} \Gamma C^* \left[1 - \left(\frac{P_e}{P_c}\right)^{(\gamma-1)/\gamma}\right]^{\frac{1}{2}}} \quad (4.36)$$

where C^* is the characteristic velocity defined in Eq. (4.33) and

$$\Gamma = [2/(\gamma+1)]^{(\gamma+1)/2(\gamma-1)} \sqrt{\gamma}. \quad (4.37)$$

The thrust coefficient is expressed as:

$$C_F = \frac{T}{P_c A^*} \quad (4.38)$$

$$M_e^2 = \frac{2}{\gamma-1} \left[\left(\frac{P_c}{P_e}\right)^{(\gamma-1)/\gamma} - 1\right] \quad (4.39)$$

If the effective exhaust velocity is expressed as C, it is defined by:

$$C = \frac{T}{\dot{m}} \quad (4.40)$$

$$C = v_e + [(P_e - P_o)/\dot{m}] A_e \quad (4.41)$$

$$C = v_e \left[1 + \frac{(P_e - P_o)A_e}{v_e(\rho_e v_e A_e)}\right] \quad (4.42)$$

or

$$C = v_e \left[1 + \frac{1}{\gamma M_e^2}\left(1 - \frac{P_o}{P_e}\right)\right] \quad (4.43)$$

Take the value of Eq. (4.31), (4.36), (4.38), (4.40), and (4.43) and solving for C_F, we get:

$$C_F = \sqrt{\frac{2\gamma}{\gamma-1}} \Gamma \left[1 - \left(\frac{P_e}{P_c}\right)^{(\gamma-1)/\gamma}\right]^{\frac{1}{2}} \times \left[1 + \frac{\gamma-1}{2\gamma}\left(\frac{P_c}{P_e}\right)^{1/\gamma} \frac{\{(P_e/P_c)-(P_o/P_c)\}}{\{1-(P_e/P_c)^{(\gamma-1)/\gamma}\}}\right] \quad (4.44)$$

and

$$C = C_F . C^* \quad (4.45)$$

Hence, the thrust coefficient is mainly depend upon the design choice of (P_c/P_e) (it is determined by area ratio) and the combustion chamber pressure and altitude.

4.4 Effect of Area Variation on Flow Properties inside the Nozzle

As we know that the gas flow through the entire nozzle is an isentropic process or flow. Let us consider a variable area duct and determined the effect of change in area, *(A)*, on the velocity *(V)*, and the pressure *(P)*. (Oates, 1997)

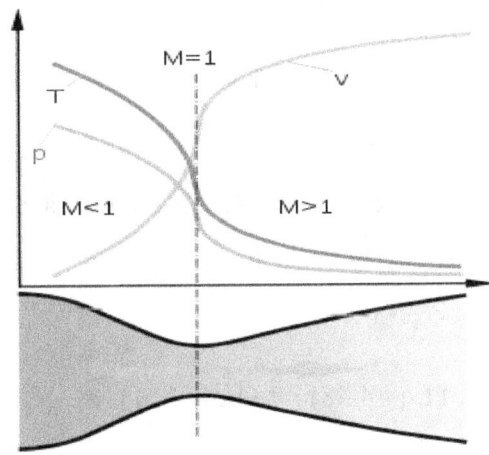

Figure 4.2: Effect of temperature, pressure, velocity inside the nozzle

According to the Bernoulli's equation:

$$\frac{dP}{\rho} + d\left(\frac{V^2}{2}\right) = 0 \tag{4.46}$$

$$dP = -\rho V \cdot dV \tag{4.47}$$

dividing by ρV^2, we get:

$$\frac{dP}{\rho V^2} = -\frac{dV}{V} \tag{4.48}$$

Differential form of the continuity equation is expressed as:

$$\frac{dA}{A} = -\frac{dV}{V} - \frac{d\rho}{\rho} \tag{4.49}$$

Substitute the value of Eq. (4.48) into Eq. (4.49), we get:

$$\frac{dA}{A} = \frac{dP}{\rho V^2}\left[1 - \frac{V^2}{dP/d\rho}\right] \tag{4.50}$$

For isentropic process, put $a^2 = dP/d\rho$ into Eq. (4.50), we get:

$$\frac{dA}{A} = \frac{dP}{\rho V^2}\left[1 - \frac{V^2}{a^2}\right] = \frac{dP}{\rho V^2}[1 - M^2] \tag{4.51}$$

From the above equation, we see that for $M < 1$, an area change causes a pressure change of the same sign and for $M > 1$, an area change causes a pressure change of opposite sign.

Again, substitute the value of Eq. (4.48) into Eq. (4.51), we get:

$$\frac{dA}{A} = -\frac{dV}{V}[1 - M^2] \tag{4.52}$$

From the above equation, we see that for $M < 1$, an area change causes a velocity change of opposite change and for $M > 1$, an area change causes a velocity change of same sign. (Orr, 1997)

5. SPACE SHUTTLE SOLID ROCKET BOOSTERS (SRBs)

Ships and sails proper for the heavenly air should be fashioned. Then there will also be people, who do not shrink from the dreary vastness of space.

The principle of the Shuttle Solid Rocket Boosters make viable the launch of the world's most developed spacecraft, the Space Shuttle. The Shuttle's propulsion system constitutes three primary components: two solid rocket booster (SRBs), the external tank (ET), and the space shuttle main engines (SSMEs). Both the external tank and the space shuttle's main engines are dependent on liquid propellants. Only the solid rocket boosters are dependent on solid propellant. Before targetting particularly on the solid rocket boosters, a short discussion of a Shuttle launch order might first be useful in understanding the function of the solid rocket boosters perform over liftoff.

At the time of launch, firstly the Shuttle's main engines, fueled as from the external tank, are ignited. When those three engines have obtained permissible thrust levels, a computer-created signal is transfered by the shuttle to fire up the solid socket boosters. The weight of individual solid rocket booster is just about 1,300,000 lb (590,000 kg) at launch. The two solid rocket boosters create almost 60% of the total lift-off mass. The weight of propellant for each solid rocket motor is just about 1,100,000 lb (500,000 kg). The neutral weight of individual solid rocket booster is just about 200,000 lb (91,000 kg). When the two SRBs are generating the proper thrust-to-weight ratio, small explosives (initiators) placed on the solid rocket boosters are to release eight hold-down bolts, release the Space Shuttle for liftoff. Just about two minutes after ingition, at an altitude of nearly 150,000 feet, the two solid boosters have exhausted their proprllant. Both solid rocket booster generates a liftoff thrust of nearly 2,800,000 lbf (12 MN) at sea level, expanding concisely following liftoff to approximately 3,100,000 lbf (14 MN). After that solid rocket booster detachment motors that are mounted on the nose and the nozzle (four on the nose and four on the nozzle) are then fired to detach the SRBs from external tank (ET). The small booster detachment motors transport them securely away from the Space Shuttle. The boosters finally originate a descend flight condition. At a fixed altitude, parachutes are disposed to slow down them for a safe landing in the ocean.

Simultaneously, the orbiter and external tank keep on to climb, utilizing the thrust of the liquid propellant controlling engines. Just about eight minutes after launch, and even short of orbital velocity, the three space shuttle main engines are closedown, and the external tank is detached from the orbiter. Once orbit is attained, particular liquid propellant orbital navigating engines are operated to keeping on the mission. (Turner, 2009)

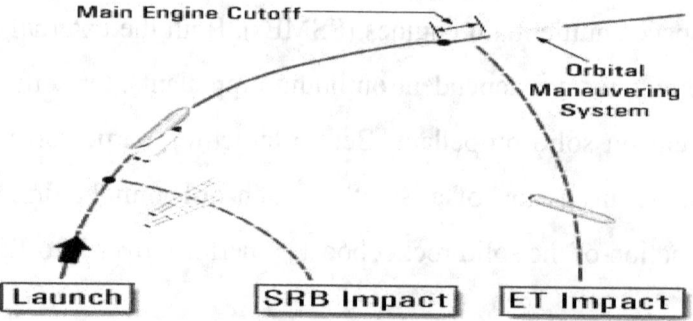

Figure 5.1: Space Shuttle launch profile

5.1 Solid Rocket Booster (SRB) Structural Design

The solid rocket boosters are very large solid propellant motors usually maneuvered and the first constructed for reprocess. Each solid rocket booster is 45.46 m long and 3.71 m in diameter. When we examine a nearer look at a solid rocket booster, it uncovers that structurally, it belongs of four particular sections of solid propellant vessels. These vessels are vertically assembled, with a nose cone on top. The nose cone area includes several electronic machines also parachutes for restoration at sea. The solid rocket booster nozzle also contains a thrust vector control system (7° gimbaling) to maintain controlling during boost climb.

All solid rocket booster section is accurately compared with the outer sections at the time of fabrication to decrease any potential thtrust instability. This "thrust-matching" is done by loading a quantity of propellant generated at the specific same time, under the specific same enviromental conditions into all of the four sections. This process is called "identical-batch".

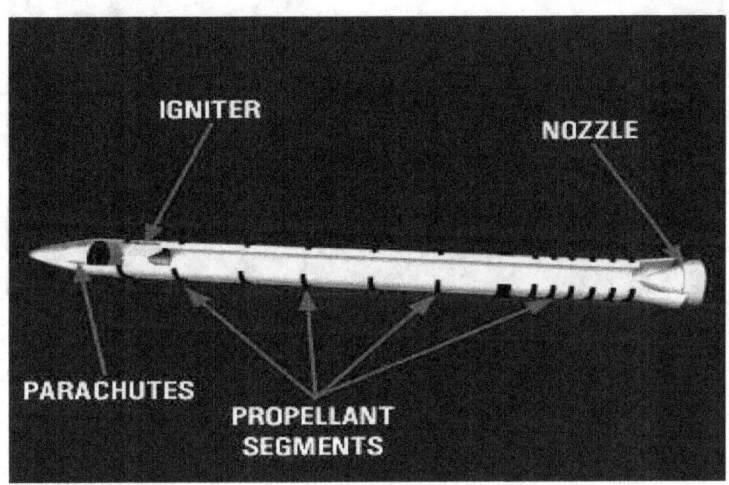

Figure 5.2: Solid rocket booster structural design

The sectioned-casing arrangement confirms ultimate flexibility in assembly and simplicity of transportation and management. All propellant-loaded sections is transported to the launch site on a heavy-duty rail car with a particularly assembled cover for on-site assembly.

5.2 Solid Rocket Booster Propellant Ingnition

Behind several arrangements pre-launch conditions have been well performed, the SRB ignition arrangement initiates while an SRB-mounted explosive machine accepts a fire instruction. These detonators fire over a narrow wall boundary seal decrease a flame tunnel. This fire up a booster charge, that is situated in the safe and arm device.

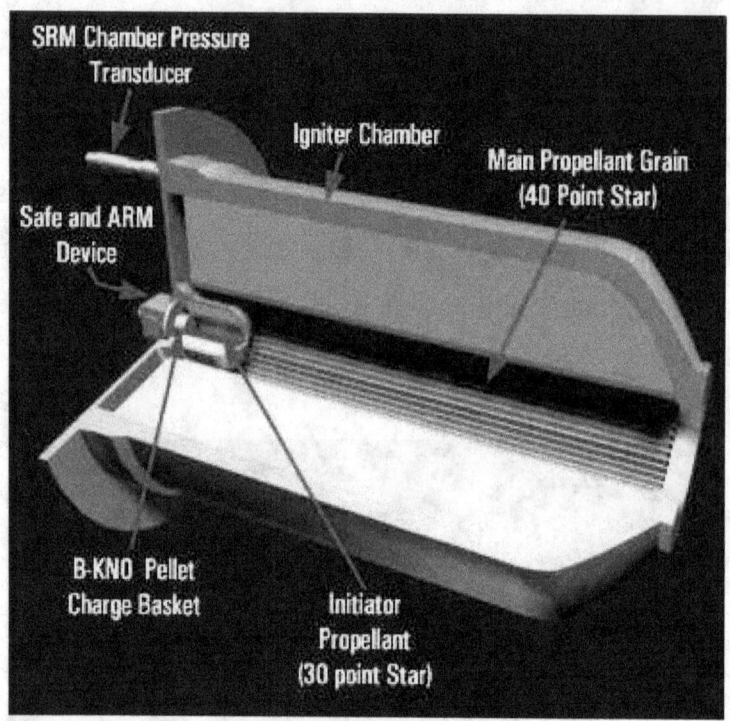

Figure 5.3: Solid Rocket Booster Propellant Ignition

The booster charge, successively, fire up the propellant in the igniter initiator, and combustion outputs of that propellant fire up the solid rocket motor initiator, that detonates a 148-foot flame the section of the rocket to fire up its propellant.

5.3 Solid Rocket Booster Propellant Features

The propellant consumed in the solid rocket booster is an 11-point star-shaped assembly in the forward motor section and a cone structured assembly in all of the afterward sections. This arrangement produces extreme thrust at ignition and as a result decreases the thrust just about 50 seconds after liftoff to avoid emphasising the vehicle as it manuevers over its highest dynamic pressure flight stage.

By their propellant exhausted after nearly two minutes igniting time, and at an altitude of just about 25 nautical miles, the SRBs are disconnected from the external tank.

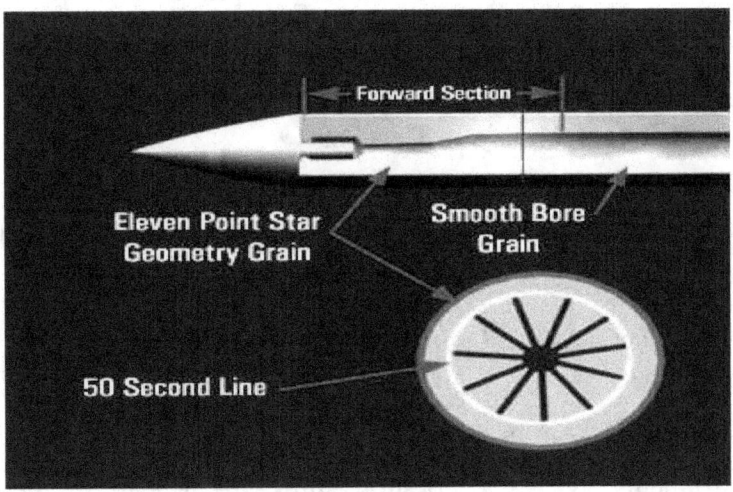

Figure 5.4: Solid rocket booster propellant features

5.4 Solid Rocket Boosters Seperation

Eight booster separation motor of both solid rocket booster generates the thrust for 1.02 seconds as disconnected from the external tank. That appears after the solid rocket booster separation arrangement software identifieds both solid rocket boosters chamber pressures under 50 psi, and that the vehicle speed is internally determined conditions.

Figure 5.5: Solid rocket boosters seperation

Those conditions check that neither solid rocket booster is even burning at separation, and that they would not strike the external tank later on separation. Solid rocket booster separation is commonly achieved mechanically, even though the flight crew can manually operated the separation if expected.

5.5 Solid Rocket Boosters Reuse and Recovery

The solid rocket boosters are slow down in their fall while at an height of nearly 3 miles, a small drogue parachute, and after that three main parachutes are arranged. That assures a reliable impact nearly 140miles out into the Atlantic Ocean.

When solid rocket boosters impact into the ocean, and guided by solid rocket booster positioning instruments, NASA restoration ships are headed to the boosters. The restoration crew recollects the solid rocket boosters, the nozzles are sealed up. The solid rocket boosters are then pulled back to the launch site, where all the portions are knock down and cleaned to restrict salt-water corrosion. The motor sections, igniter and nozzle are tarnsported back to the manufacturer, for renewal and future reuse. (Wikipedia)

5.6 Summary

At the time of launch, the two solid rockets exhausts more than 10 tons of fuel every second and generates 44 million hp. They produce the leading thrust to take up the Space Shuttle off the pad and up to an height of about 25 nautical miles. The benifits provided by these, or any solid-propellant rocket, are that they integrate a high level of depedability with an functionnal simplicity not accessible on other rocket designs. As the earlier days of invention and examination, solid propellant rockets have developed from simple gunpowder instruments into large vehicles experienced of wander in outer space.

6. INVISCID AND COMPRESSIBLE FLOW THROUGH A CONVERGING-DIVERGING NOZZLE

The shape of the heaven is of necessity spherical; for that is the shape most appropriate to its substance and also by nature primary.

In this chapter the obecjtive is to illustrate the setup and solution of an axisymmetric fluid flow through a nozzle.

The flow through a converging-diverging nozzle is one of the benchmark problems used for modeling the compressible flow through computational fluid dynamics. Occurrence of shock in the flow field displays one of the most prominent effects of compressibility over a fluid flow. Accurate shock predication is a challenge to the CFD fraternity. In order to resolve the high pressure gradients we need to use some special numerical schemes along with fine grid . These are the following steps for modeling the copmressible flow through CFD.

- ✓ Read the mesh file in FLUENT from your working folder.
- ✓ Check the grid for dimensions and quality.
- ✓ Change the material properties.
- ✓ Perform inviscid calculations.
- ✓ Compare the results for different models.

6.1 Description

Let us consider a converging-diverging nozzle length (L) of 0.6 m. The inlet radius (r1) is 0.1 m and the outlet radius (r2) is 0.12 m. The ratio of throat area to the inlet area is 0.5625. The pressure differnce across the nozzle is 0.12 MPa. Figure (6.1) shows longitudnal section of converging-diverging, symmetric about the axis.

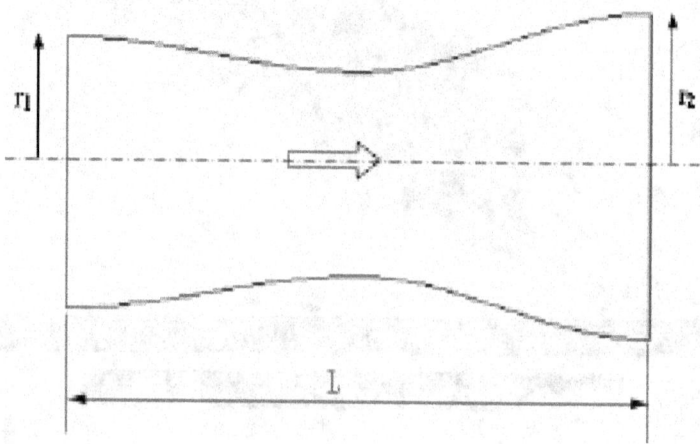

Figure 6.1: Problem schematic

6.2 Setup and Solution for Viscous Flow

Step 1 : Grid

1. Read the grid file nozzle.msh.

    FLUENT will read the mesh file and report the progress in the console.

2. Check the grid.

    This procedure checks the integrity of the mesh. Make sure the reported minimum volume is a positive number.

3. Check the scale of the grid.

Check the domain extents to see if they correspond to the actual physical dimensions. Otherwise the grid has to be scaled with proper units.

4. Display the grid.

The grid adjacent to the walls is finer as compared to that in the central region. The purpose for such fine mesh is to capture sharp gradients near the walls correctly.

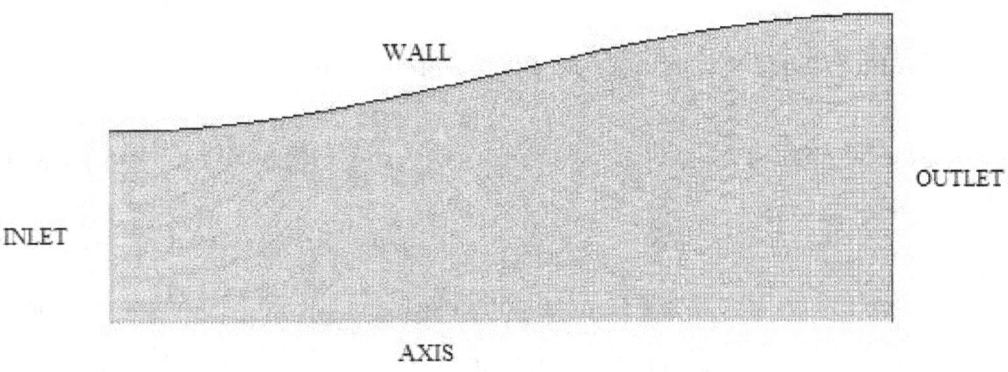

Figure 6.2: Grid Display

Step 2: Models

In this step, we define solver, viscous, and material properties.

In solver: Select axisymmetric form the space list.

In viscous: Select k-epsilon (2 eqn) from the model list.

In materials: Select ideal-gas from the density drop-down list.

Energy equation will get enabled as soon as ideal gas density formulation is used.

Step 3: Operating conditions

Enter 0 for operating pressure.

For compressible flow, it is recommended to set the operating pressure to zero to minimize the errors due to pressure fluctuation.

Step 4: Boundary conditions

Set the boundary conditions for inlet as pressure inlet.

Enter 220000 Pa for Gauge total pressure and 210000 Pa for supersonic/initial guage pressure.

Enter 1% and 0.2 m for turbulance intesity and hydraulic diameter respectively. For higher reynolds number flow, turbulance intensity is in the range of 1-5%. In this case set it to 1% as the diameter of inlet is 0.2 m. Set the total temperature to 300K.

Set boundary condition for outlet as pressure outlet.

Enter 100000 Pa for Gauge pressure, 1% for turbulance intensity, and 0.24 m for hydraulic diameter corresponding to the diameter of the outlet. The outlet is assumed to open in the atmosphere. So the outlet pressure is set approximately equal to the atmospheric presure. Set temperature to 300K for backflow total teperature.

Step 5: Solution

Set the default setting for solution controls and after that initialize the flow for inlet. It will update values of all the variables based on the boundary conditons at inlet. Enable the plotting of residuals during calculations.

Start the calculation by requesting 2000 iterations. The solution converges in about 1166 iterations with specified convergence criteria. The residuals plot is shown in figure (6.3).

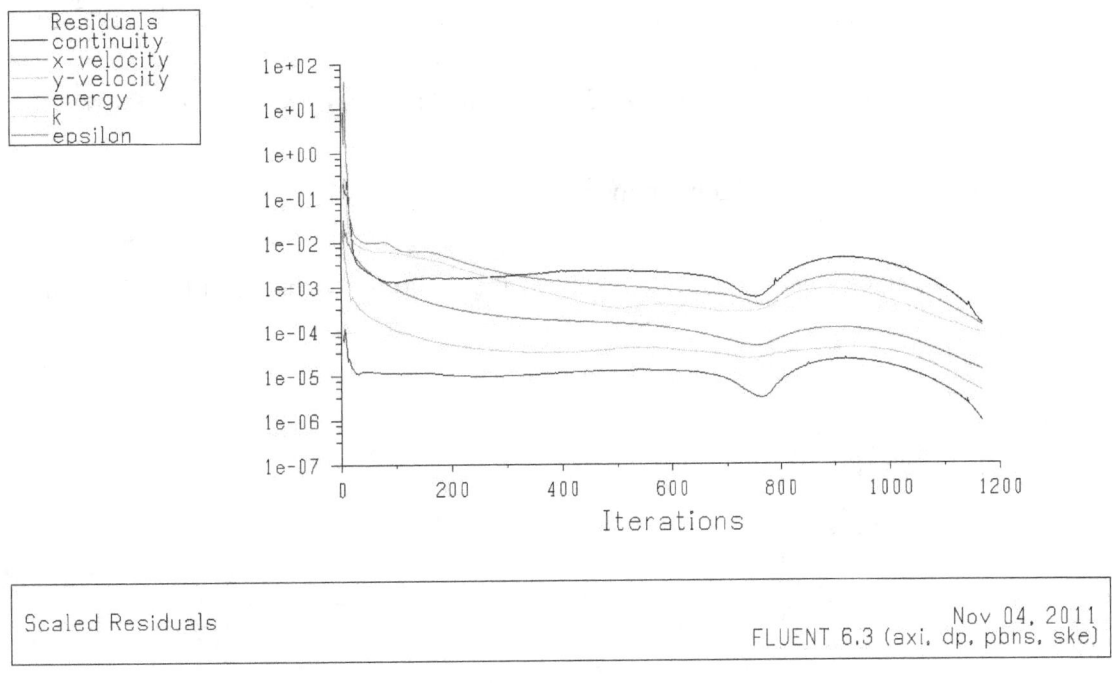

Figure 6.3: Scaled Residuals

Display contours of static pressure for Viscous Flow:

Select pressure and static pressure from the contours to display filled contours of static pressure.

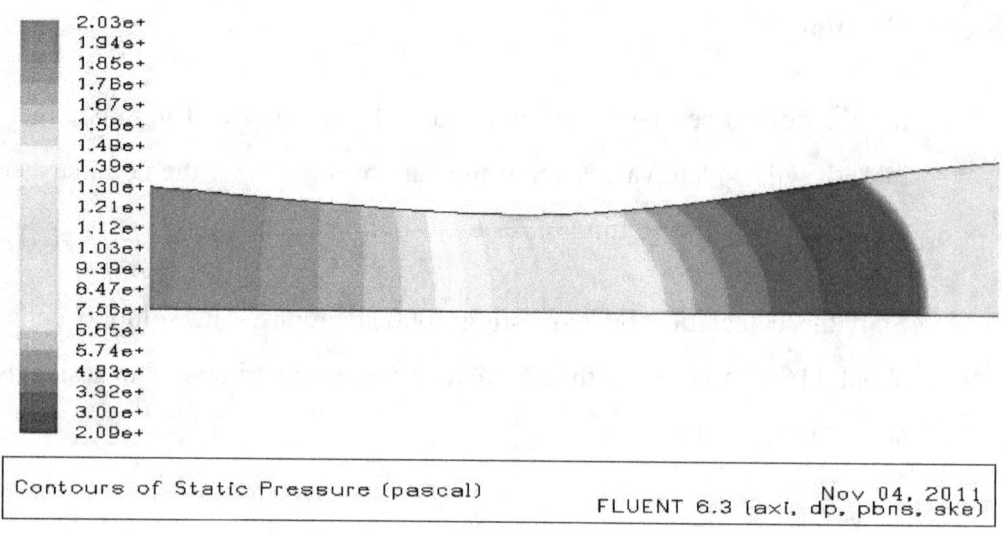

Figure 6.4: Contours of static pressure (Viscous Flow)

Display contours of Mach number:

Select velocity and Mach number from the contours to display filled contours of Mach number.

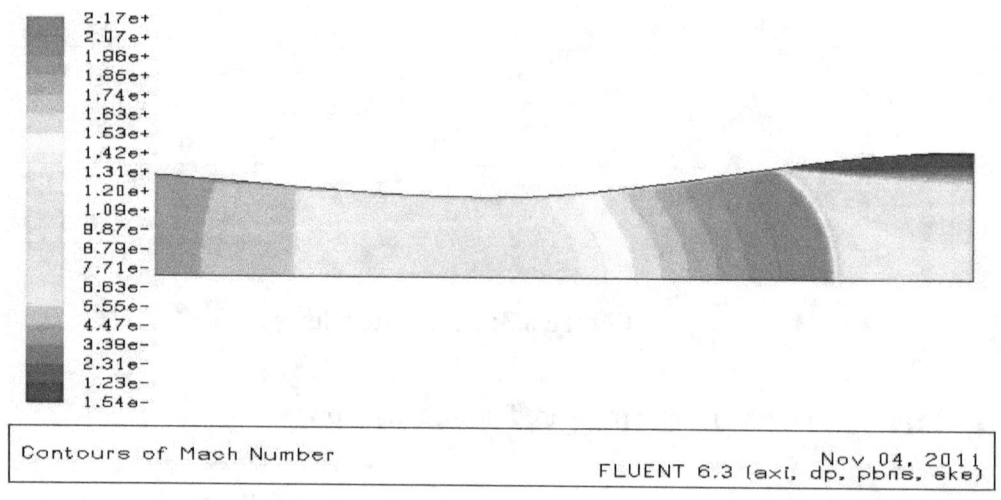

Figure 6.5: Contours of Mach number (Viscous Flow)

Display contours of static temperature:

Select tempersture and static temperature from the contours to display filled contours of static temperature.

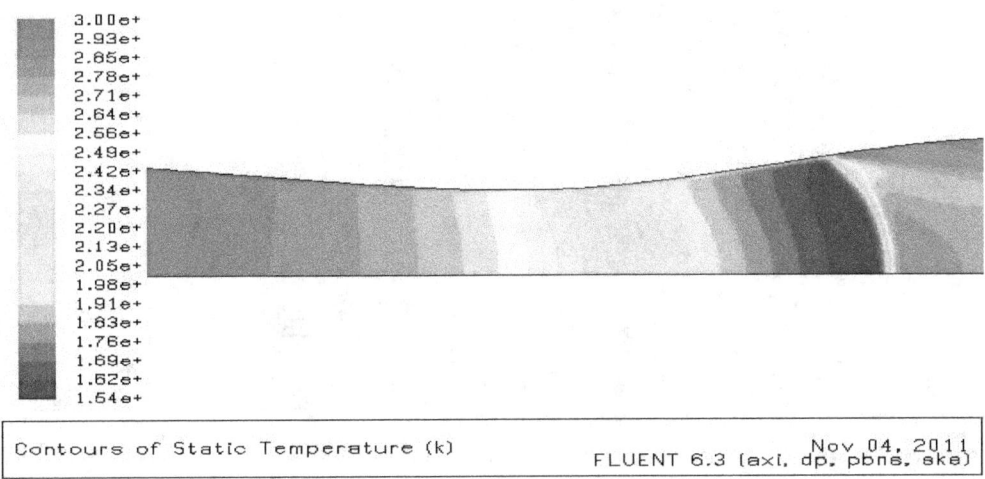

Figure 6.6: Contours of static temperature (Viscous Flow)

Display velocity vectors:

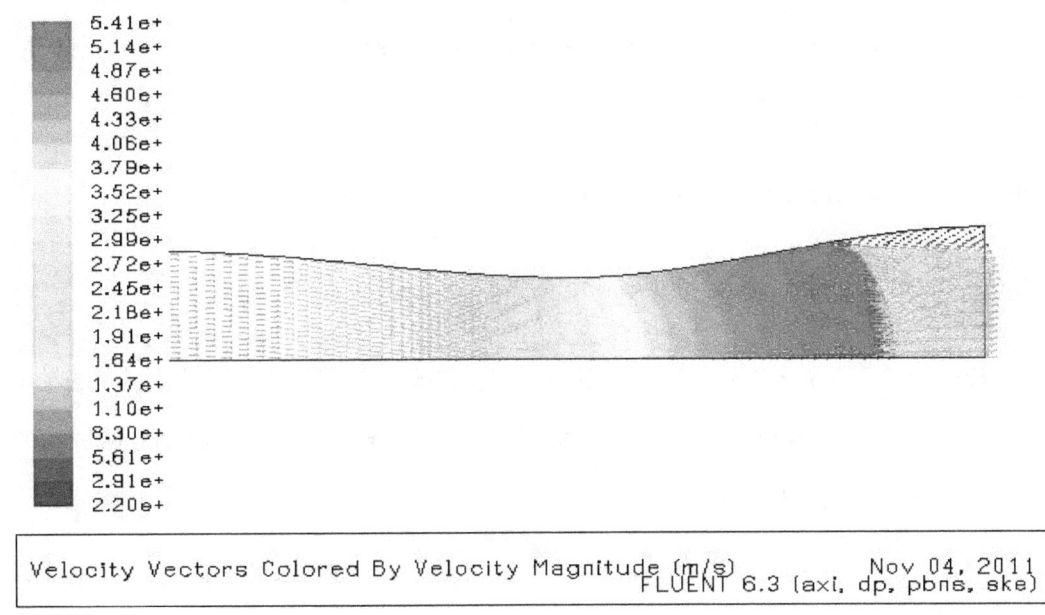

Figure 6.7: Velocity vectors (Viscous Flow)

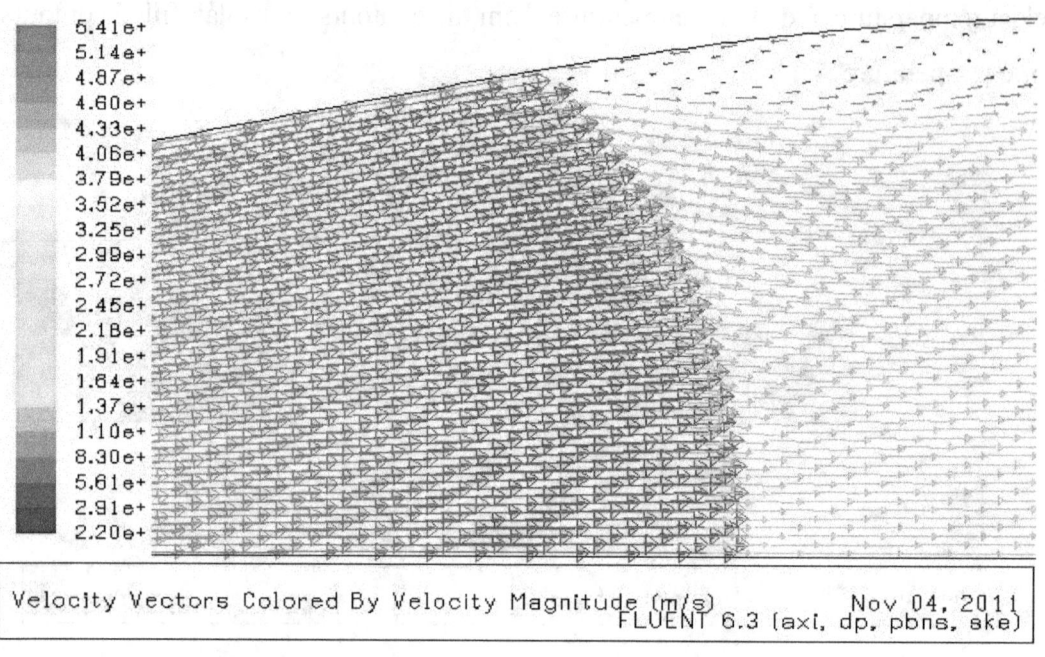

Figure 6.8: Magnified view of velocity vectors

Observe the reversed flow at the top end of the outlet. The reason for this is the shock after which the pressure gradient becomes adverse. This causes flow seperation and a vortex is formed. The pressure outlet intersects the vortex and result in a reversed flow. We can extend the domain to avoid reverse flow.

6.3 Setup and Solution for Inviscid Flow

Enable inviscid viscous model and initialize the flow.

Start the calculation by requesting 1000 iterations. The solution converges in about 450 iterations, with the default convergence criteria. The resiuals plot is shown in figure (6.7):

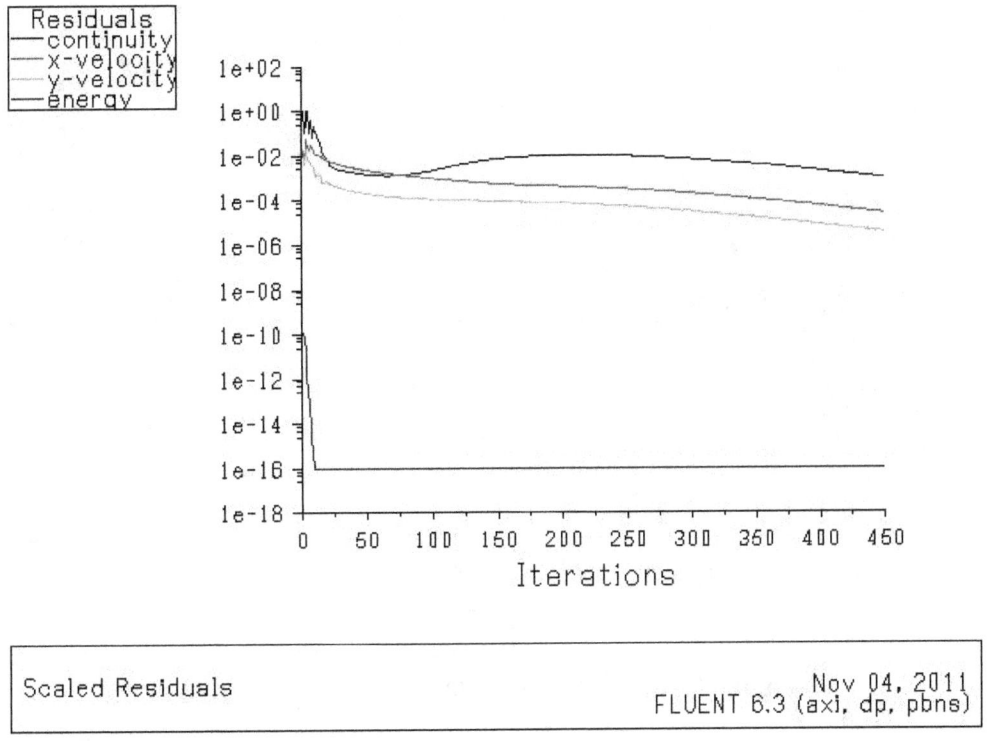

Figure 6.9: Scaled residuals

Display contours of static pressure:

Select pressure and static pressure from the contours to display filled contours of static pressure.

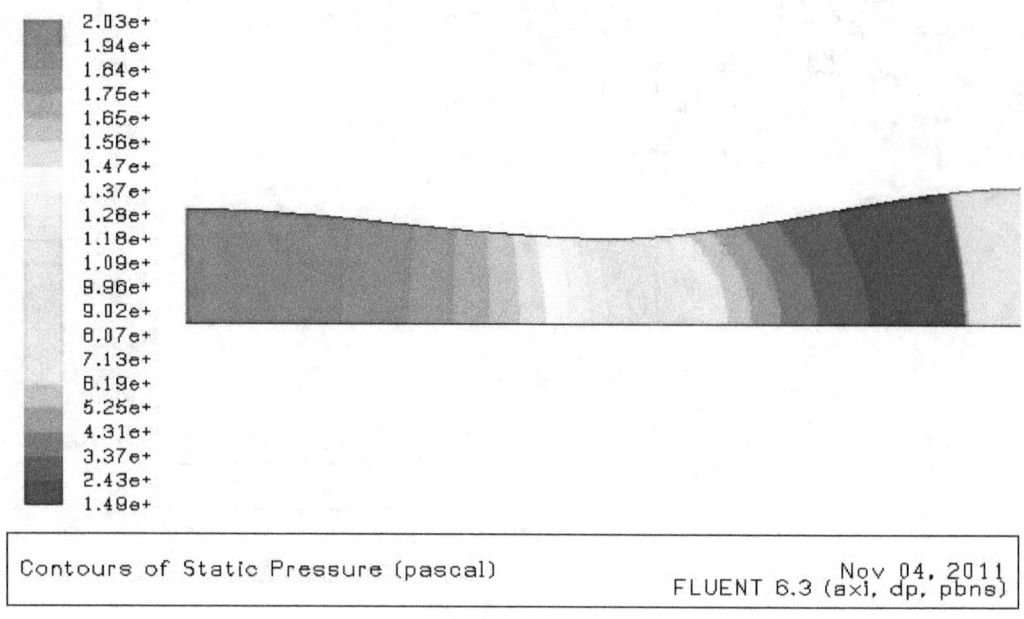

Figure 6.10: Contours of static pressure (Inviscid Flow)

Display contours of Mach number:

Select velocity and Mach number from the contours to display filled contours of Mach number.

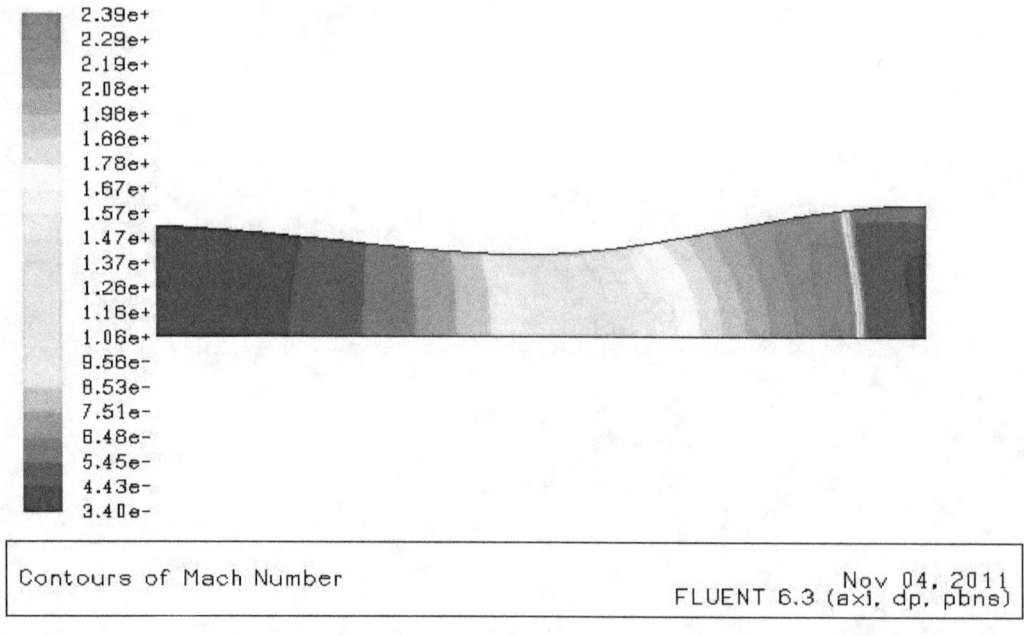

Figure 6.11: Contours of Mach number (Inviscid Flow)

Display contours of static temperature:

Select temperature and static temperature from the contours to display filled contours of static temperature.

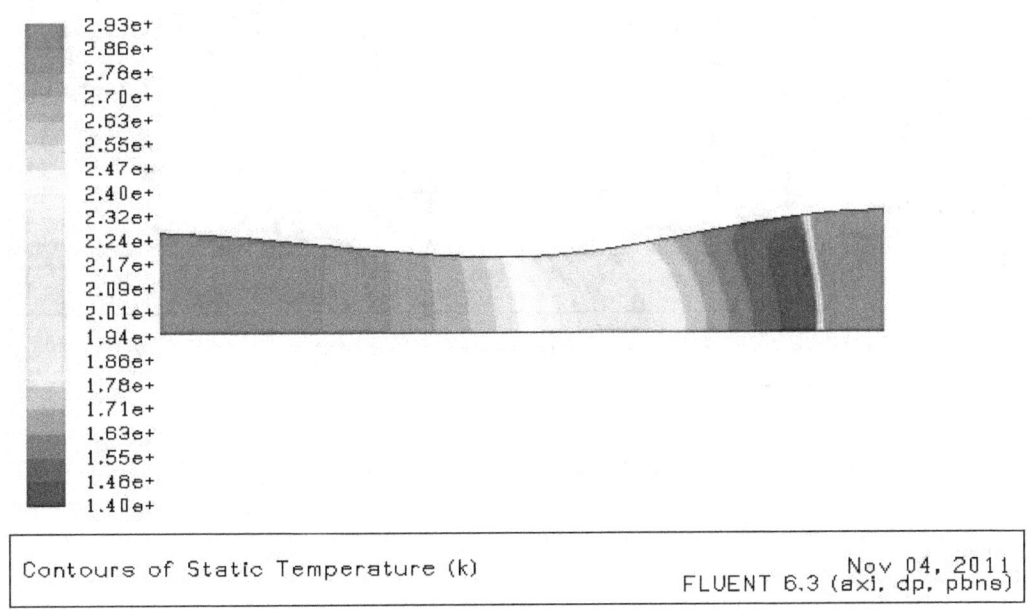

Figure 6.12: Contours of static temperature (Inviscid Flow)

Display vectors:

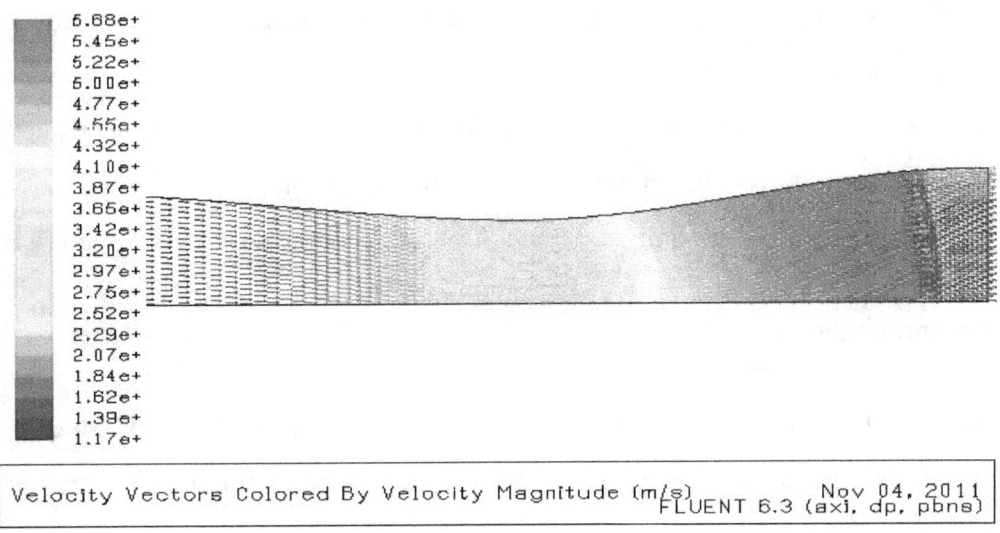

Figure 6.13: Velocity vectors (Inviscid Flow)

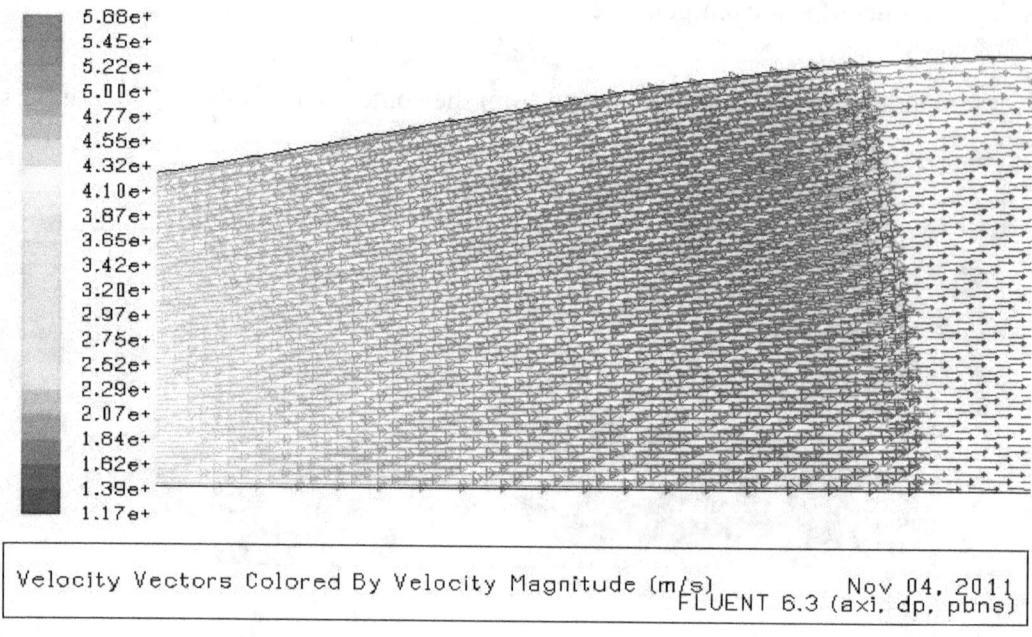

Figure 6.14: Magnified view of velocity vectors (Inviscid Flow)

6.4 Results

- ✓ The inviscid flow solution shows a straight shock instead of curved shock. In viscous flow, velocity is less near the nozzle. So, the shock takes place near the wall before it takes place in the main flow. Hence, the location near the wall is upstream to the shock location near the axis.
- ✓ Flow separation does not occur in inviscid flows. Hence, no vortex formed near the outlet.
- ✓ The viscosity accounts to loss in momentum. As this loss is not considered in inviscid flow, a higher maximum Mach number is obtained.
- ✓ Inviscid flow gives a quick estimate of shock location and flow characteristics.

6.5 Recommendation

As shown in FLUENT analysis in above figures, the flow is slowing down just before exiting the nozzle due to flow speration and vortex is formed. So, the divergent part of the nozzle nedds to be shortened. Hence, a shorter convergent-divergent profile is needed.

REFERENCES

Turner, Martin J. L., 2009: Rocket and Spacecraft Propulsion: Principles of Rocket Propulsion, pp. 10-17, Solid Propellant Rocket Motors, pp. 109-124, Praxis, Chichester, UK.

Oates, Gordon C., 1997: Aerothermodynamics of Gas Turbine and Rocket Propulsion: Thermodynamics and Quasi-One Dimensional Fluid Flows, The Laws of Thermodynamics, pp. 21-26, The Nozzle Flow Equation, pp. 52-53, American Institute of Areonautics and astronautics, Inc., Reston, Virginia.

Dr. Rogers, Lucy, 2008: It's Only Rocket Science: Chemical Rocket Propulsion, pp. 116-127, Electrical Propulsion, p. 127, Nuclear Propulsion, p. 131, Solar Thermal Propulsion, Other forms of Propulsion, pp. 131-132, Springer Science + Business Media, New York, USA.

Kuentzmann, P., R. J., 2002: Introduction to Solid Rocket Propulsion: Fundamental Aspects of Solid Propellant Rockets, Internal Aerodynamics in Solid Rocket Propulsion, France.

Sutton, George P., and Biblarz, Oscar 2001: Rocket Propulsion Elements: Definitions and Fundamentals, pp. 27-38, Nozzle Theory and Thermodynamics Relations, pp. 45-53, John Wiley and Sons, New York, USA.

Orr, Graham, 2009: Estimation and Analysis of Quasi-ID Solid Rocket Motor, Harvey Mudd College.

Url-1 <*http://www.aerospacelectures.co.cc*>, accessed at 01.11.2011.

Url-2 <*http://www.howstuffworks.com*>, accessed at 01.09.2011.

Dr. Ugur Guven is an Aerospace Engineer as well as a Nuclear Engineer. His area of interest is Nuclear Space Propulsion and Interstellar Propulsion Methods along with Utilization of CFD in analyzing Nuclear Reactors. In nuclear power plant engineering, he is interested in Generation IV Reactors, Gas Core Reactors and Helium Cooled Reactors. Dr. GUVEN is also TRIGA Mark II Certified Nuclear Plant Operator.

Dr Ugur Guven has published many papers in the field of interstellar space exploration, mission to alpha centauri, nuclear propulsion techniques for spacecraft, Utilization of Nuclear Energy on the Moon, Helium 3 Mining on the Moon for Fusion Technologies, as well as on a Mission to Mars using Nuclear Propulsion Techniques. He has also written over 30 opinion articles on the future of space technology and nuclear technology. Dr. Ugur Guven has published one book on *"Nuclear Propulsion Techniques for Spacecraft"* ISBN: 978-3-8473-2269-6, as well as another book on *"Nuclear Power Plant Design Using Gas Cooled Nuclear Reactors"*. ISBN: 978-3-8484-2176-3.

Dr. Guven has taught wide range of courses from Aerodynamics, Aeroelasticity, Space Sciences, Plasma Physics, Nuclear Power Generation, Nuclear Physics, Spacecraft Dynamics, Satellite Communications, CFD, Rocket Propulsion, and Advanced Numerical Methods for Engineers, FORTRAN Programming for Engineers, MATLAB Techniques for Advanced Numerical Methods, and Introduction to Aerospace Engineering. Dr. Ugur Guven is currently involved as a coordinator in a Nanosatellite project.

Dr. Guven has traveled to over 25 countries and he has worked in USA, Turkey, India, Mauritius, Switzerland and other locations. Dr. Guven has received a commendation for "The Best Teacher of the
Year Award". His lectures and seminars are downloaded by students, teachers and engineers worldwide from his various engineering websites. Dr Guven lecture sites include Aerospace Lectures, Space Lectures, CFD Lectures, IT Lectures and Nuclear Lectures. Dr. Ugur Guven also works in improving educational and pedagogical techniques in higher education. He is also the Chief Editor of Scholar Journal for Young Scholars and Researchers (ISSN: 2164-5132), which he has created and published for the benefit of engineering students and graduates. Dr. Guven travels frequently all over the world to give free seminars on the future of space technology and nuclear technology.

Gurunadh Velidi is a Doctoral Research fellow in the field of Nuclear Space Propulsion, and currently working on various aspects of deep space launch vehicle design. He also published 5 Journal papers and 24 International conference proceedings and 2 National Conference Proceedings in the area of Advanced Nuclear Power generation Methods, Nuclear Reactors for Space Applications, Mathematical Modeling of Nuclear reactions in various Environments and Interstellar travel. Gurunadh Velidi is an author for two books, "Nuclear Power Plant Design using Gas Cooled Modular Helium Cooled Reactors", and Exergy Analysis on Combined Cycle Power Plants". Gurunadh Velidi recently attended the International Astronautical Congress 2012 in Napoli, Italy and presented his recent out comes from his research.

Gurunadh Velidi is basically a Mechanical Engineer with the M.Tech in Energy Systems with Nuclear Engineering Specialization, he had worked on various Academic projects like "UPESSAT"(University Satellite Program), Exhaust Emissions testing with different roots of Biodiesel and ethanol based blending fuels with Thermal Engineering and Heavy Engines laboratory in Pondicherry University and also a Academic coordinator to M.TECH Computational Fluid Dynamics Program in University of petroleum and Energy Studies. He also worked as a Energy Ambassador for the Pondicherry University Energy Conservation Program, and worked a Chair person to SAE Student Chapter. His Current Research Interests are Advanced Nuclear Engineering for Space Applications, generation IV Reactor systems and Interstellar Travel, Modeling of Controlled fission methods for Space and remote power generation models.

www.ingramcontent.com/pod-product-compliance
Lightning Source LLC
Chambersburg PA
CBHW060436220526

45465CB00008B/3156